ELECTRICAL TRANSMISSION —AND— DISTRIBUTION

S RAMA SUBBANNA
B LOVESWARA RAO

INDIA • SINGAPORE • MALAYSIA

Notion Press

Old No. 38, New No. 6
McNichols Road, Chetpet
Chennai - 600 031

First Published by Notion Press 2019
Copyright © S Rama Subbanna, B Loveswara Rao 2019
All Rights Reserved.

ISBN 978-1-64650-590-6

This book has been published with all efforts taken to make the material error-free after the consent of the author. However, the author and the publisher do not assume and hereby disclaim any liability to any party for any loss, damage, or disruption caused by errors or omissions, whether such errors or omissions result from negligence, accident, or any other cause.

While every effort has been made to avoid any mistake or omission, this publication is being sold on the condition and understanding that neither the author nor the publishers or printers would be liable in any manner to any person by reason of any mistake or omission in this publication or for any action taken or omitted to be taken or advice rendered or accepted on the basis of this work. For any defect in printing or binding the publishers will be liable only to replace the defective copy by another copy of this work then available.

Dedicated to my Mother
Late Smt. Grace Somarpu

– **Dr. Rama Subbanna Somarpu**

Contents

Preface 11

Chapter 1. Electrical Design of Overhead Transmission Lines 13
    1.1 Introduction 13
    1.2 Typical Power Supply Systems 13
    1.3 Why Higher Voltages for
          Transmission of Electric Power? 16
    1.4 Conductors 19
          1.4.1 Conductor Materials 20
          1.4.2 Types of Conductors 22
          1.4.3 Construction of a Stranded Conductor 25
    1.5 Transmission Line Parameters 27
          1.5.1 Resistance 28
          1.5.2 Inductance 29
          1.5.3 Capacitance 29
          1.5.4 Conductance 30
    1.6 Skin Effect 30
    1.7 Proximity Effect 32
    1.8 Concept of GMR & GMD 33
    1.9 Line Inductance 37
          1.9.1 Flux Linkages of a Conductor
                Due to Internal Flux 37
          1.9.2 Flux Linkages of a Conductor
                Due to External Flux 40
          1.9.3 Inductance of $1-\phi$ Two Wire Line 41

1.10 Flux Linkages of One Conductor in a
Group of Conductors 43
1.11 Inductance of 3ϕ Single Circuit Overhead
Transmission Law 45
1.12 Inductance of 3ϕ Line with Double Circuit 54
    1.12.1 Inductance of a 3ϕ Double
           Circuit Line with Symmetrical Spacing 54
    1.12.2 Inductance of a 3ϕ Double Circuit with
           Unsymmetrical Spacing But Transposed 56
1.13 Capacitance of a Transmission Line 59
    1.13.1 Potential at a Charged Single Conductor 60
    1.13.2 Potential at a Charged Conductor in a Group of
           Charged Conductors 60
    1.13.3 Capacitance of a 1ϕ2 Wire Overhead Line ... 62
1.14 Effect of Earth on the Capacitance of 1ϕ OHTL 63
1.15 Capacitance of 3ϕ Single Circuit Equilateral Spacing 65
1.16 Capacitance of 3 – ϕ Double Circuit with Symmetrical
Configuration 67
1.17 Capacitance of 3 – ϕ Single Circuit Overhead Line
(Unsymmetrical Configuration) 69
1.18 Capacitance of 3 – ϕ Double Circuit with
Unsymmetrical Configuration 71

**Chapter 2. Performance of Short and Medium
Length Transmission Lines 81**
2.1 Introduction 81
2.2 Classification of Transmission Lines 82
2.3 Regulation and Efficiency of a Transmission Line ... 83
2.4 Short Transmission Lines 83
2.5 Generalized Circuit Parameters 86
2.6 Medium Transmission Lines 88
    2.6.1 Nominal – T Method: 89
    2.6.2 Nominal – Π Method: 94

Contents | 7

Chapter 3.	Performance of Long Transmission Lines 125

    3.1 Introduction 125
    3.2 Analysis of Long Transmission
          Lines By Rigorous Solution Method 126
    3.3 Calculation of A B C D Constants for
          Long Transmission Lines 129
    3.4 Evaluation of A B C D Constants 130
    3.5 Surge Impedance 131
    3.6 Surge Impedance Loading (SIL) 133
    3.7 Wave Length & Velocity of Propagation of Wave . . . 133

Chapter 4. Power System Transients. 141
    4.1 Introduction 141
    4.2 Types of System Transients. 141
          4.2.1 Surge Phenomena 142
          4.2.2 Short-Circuit Phenomena 142
          4.2.3 Transient Stability 143
    4.3 Attenuation & Distortion of Travelling Waves 143
    4.4 Reflection & Refraction of Travelling Waves 144
    4.5 Termination of Transmission Line 147
          4.5.1 With Open End 147
          4.5.2 With Short Circuit End 149
          4.5.3 With T-Junction 151
    4.6 Bewley's Lattice Diagrams 154

Chapter 5. Corona 159
    5.1 Corona Phenomenon 159
    5.2 Critical Disruptive Voltage: (V_0) 161
    5.3 Critical Visual Voltage (V_v) 164
    5.4 Corona Loss: (P) 165
    5.5 Factors Affecting Corona Loss. 167
    5.6 Methods of Reducing Corona 169

Chapter 6. Mechanical Design of Overhead Transmission Lines.. 177

- 6.1 Overhead Line Insulators 177
  - Importances of an Insulator: 177
- 6.2 Properties of Insulators.........................177
- 6.3 Materials Used for Insulators Are Two Types 178
- 6.4 Types of Insulators.............................179
  - 6.4.1 Pin Type Insulator 179
  - 6.4.2 Suspension Type Insulators...............180
  - 6.4.3 Strain Type Insulator:
    (Or) Tension Insulator...................182
  - 6.4.4 Shackle Type Insulator...................182
- 6.5 Voltage Distribution Among Discs (Units) of
  Suspension Type Insulator 183
- 6.6 Methods to Improve the String Efficiency.........185
  - 6.6.1 Using Longer Cross – Arm 186
  - 6.6.2 Capacitance Grading 186
  - 6.6.3 Static Shielding / Guard Ring Method......188
- 6.7 Sag & Tension Calculations 199
- 6.8 Factors Affecting Sag..........................200
- 6.9 Calculation of Sag & Tension....................201
  - 6.9.1 With Equal Level Supports 201
  - 6.9.2 At Unequal Level of Supports.............206
- 6.10 Effect of Wind and Ice on
  the Weight of the Conductor 207
- 6.11 Sag Template..................................209

Chapter 7. Underground Cables 221

- 7.1 Introduction..................................221
- 7.2 Properties of Insulating Material for Cable 221
- 7.3 Requirements of an Underground Cable 222
- 7.4 Classification of Cables.........................223
- 7.5 Construction of a Single Core Cable..............223
- 7.6 Insulating Materials for Underground Cables......225
- 7.7 Insulation Resistance of a Cable 228

	7.8	Capacitance of a Single Core Cable............229
	7.9	Dielectric Stress in a Single Core Cable..........231
	7.10	Most Economical Size (Or Diameter) of Conductor 233
	7.11	Grading of Cables 235
		7.11.1 Capacitance Grading 236
		7.11.2 Inter Sheath Grading 239
	7.12	Capacitance of 3-Core Belted Cables 245

Chapter 8. Distribution Systems............................. 259

- 8.1 Introduction 259
- 8.2 Distribution System.............................259
  Components of Distribution System..............259
- 8.3 Classification of Distribution Systems 260
  - 8.3.1 A.C Distribution........................261
  - 8.3.2 D.C Distribution.......................262
- 8.4 Overhead versus Underground System 263
- 8.5 Connection Schemes of distribution System.......264
  - 8.5.1 Radial System 264
  - 8.5.2 Ring Main System......................265
  - 8.5.3 Inter Connected System..................266
- 8.6 Requirements of a Distribution System 267
- 8.7 Types of D.C Distributors......................268
  - 8.7.1 Distributor Fed at One End
    (Concentrated Loading) 268
  - 8.7.2 Distributor Fed at Both Ends 269
  - 8.7.3 Distributor Fed at the Centre 270
- 8.8 Distributor Fed at One End –
  Uniformly Distributed Load 274
- 8.9 Distributor Fed at Both Ends –
  Concentrated Loading 278
  - 8.9.1 Two Ends with Equal Voltages 278
  - 8.9.2 Two Ends Fed with Unequal Voltages 279
- 8.10 Distributor Fed at Both Ends – Uniform Loading ..284

        8.10.1 Distributor Fed at Both
               Ends with Equal Voltages 284
        8.10.2 Distributor Fed at Both Ends
               with Unequal Voltage....................285
   8.11 A.C Distribution System........................291
        8.11.1 Power Factor Referred to
               Receiving End Voltage...................292
        8.11.2 Power Factors Referred to
               Respective Load Voltages................293
   8.12 Stepped (or) Tapped Distributor................299

Objective Questions *303*

Preface

This book is mainly divided into two sections. The first section deals with power supply schemes, overhead transmission of electrical power, conductor materials, electrical and mechanical design aspects of transmission lines, performance of transmission lines, different phenomena that occur in the transmission system and overhead. It also covers the transmission of electric power by underground cables.

The second section deals with electrical distribution system, where D.C. and A.C. distribution system concepts, different types of D.C. distribution schemes and different solutions to solve the A.C. distribution problems are covered. The book covers the syllabi of many universities in India for a course in power transmission and distribution.

Chapter 1 describes the electrical design aspects of overhead transmission lines and provides different supply schemes, conductor materials, types of conductors, concept of GMR and GMD, calculation of inductance and capacitance for different circuit's under different configurations.

Chapter 2 deals with types of overhead transmission lines and talks specifically about short and medium transmission lines. Calculation of A B C D constants, regulation and efficiency of short and medium transmission lines.

Chapter 3 looks at the derivation of Rigorous Solution for analysing the performance of long transmission lines. Explanation of long line equations, Calculation of A B C D constants, Wave length and Velocity of propagation of waves.

Chapter 4 explains the types of power system transients, termination of transmission lines under various conditions, incident, reflected and refracted voltages and lattice diagram.

Chapter 5 deals with the phenomenon of Corona. Power loss by Corona under various weather conditions, critical and visual disruptive voltages, different methods to reduce Corona.

Chapter 6 Expounds the mechanical design aspects of overhead transmission lines. Provides the information about types of insulators in overhead transmission system, voltage distribution among the insulator discs in the string, string efficiency and insulator grading methods. Sag and tension calculations of a line at different height level of towers.

Chapter 7 deals with underground cables construction, classifications, insulating materials, dielectric stress, calculation of insulation resistance and capacitance of single and three core belted cables. Grading of the cables and solved number of problems.

Chapter 8 Describes electric distribution system, where D.C. and A.C. distribution system concepts, different types of D.C. distribution schemes and different solutions to solve the A.C. distribution problems are covered.

Every care has taken to eliminate misprints and errors but it is too much to expect that no inaccuracy/misprint/error has crept in and the author would be grateful to the readers for bringing to his notice any such errors/misprints/omissions they may come across while going through the book.

The author's shall feel satisfied if the book meets the needs of the students for whom it is meant. Suggestions and constructive criticism for the improvement of the book shall be welcomed.

The author's will like to thank the readers (both teachers and students) who sent their valuable suggestion that become the basis for the revision of the book.

Dr. S. Rama Subbanna
Dr. B. Loveswara Rao

CHAPTER 1

Electrical Design of Overhead Transmission Lines

1.1 Introduction

Transmission of electric power is done by 3-phase, 3-wire overhead lines. An A.C. transmission line has resistance, inductance and capacitance & uniformly distributed along its length. These are known as 'line constants or 'line parameters'. The performance of a transmission line depends to a considerable extent upon these constants. For instance, these constants determine whether the efficiency and voltage regulation of the line will be good or poor. Therefore a sound concept of these constants is necessary in order to make the electrical design of a transmission line a technical success.

1.2 Typical Power Supply Systems

The Large network of Conductors between the power station and the Consumers can be broadly classified into two parts i.e., transmission system and distribution system. Each part can be further sub-divided into primary transmission & secondary transmission and primary distribution and secondary distribution. Fig 1.1. Shows the layout of typical A.C. power supply system by a single line diagram. It may be noted that it is not necessary that all power schemes include all the stages shown in figure."

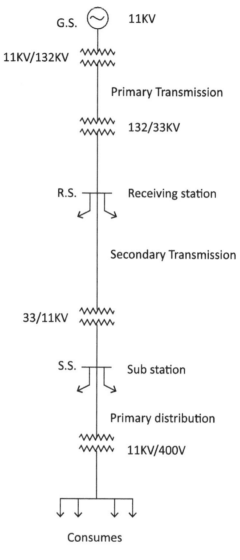

Fig. 1.1

(i) Generating station (G.S.):

In fig 1.1, G.S. represents the generating Station, where electric power is produced by 3-phase alternators operating in parallel. The usual generation voltage is 11KV or even 33KV in certain cases. For economy in the transmission of electric power, the generation voltage (i.e. 11KV) is stepped up to 132 KV or more (Depending upon the length of transmission line and amount of power to be transmitted) at the generating stations with the help of 3-phase transformers. Generally the primary transmission is carried at 66KV, 132 KV, 220KV or 400KV.

(ii) **Primary transmission:**

The electric power at 132 KV is transmitted by 3-phase, 3-wire overhead system to the outskirts of the city. This farms the primary transmission.

(iii) **Secondary transmission:**

The primary transmission line terminates at the receiving station (RS) which usually lies at the outskirts of the city. At the receiving station, the voltage is reduced to 33KV by step down transformer. From this station, electric power is transmitted at 33KV by 3-phase, 3-wire overhead system to various sub stations (SS) located at the strategic points in the city. This forms the secondary transmission.

(iv) **Primary Distribution:**

The secondary transmission line terminates at the sub-station where voltage is reduced from 33KV to 11KV. The 11KV lines run along the important road sides of the city. This forms the primary distribution. It may be noted that big consumers (having demand more than 50KW) are generally supplied power at 11KV for further handling with their own sub-stations.

(v) **Secondary Distribution:**

The electric power from Primary distribution line (11 KV) is delivered to distribution sub-station (DS). These sub-stations are located near the consumer's localities and step down the voltage to 400V, 3-phase, and 4-wire for secondary distribution. The voltage between any two phases is 400 V and between any phase and neutral is 230 V. The single phase residential lighting load is connected between any one phase and neutral, whereas 3-phase, 400 V motor loads is connected across 3-phase lines directly.

NOTE: Secondary distribution system consists of feeders, distributors and service mains.

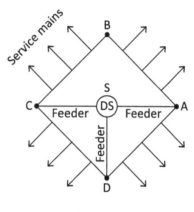

Fig. 1.2

Fig 1.2 shows the elements of low voltage distribution system. Feeders (SC or SA) radiating from the distribution sub-station (DS) supply power to the distributors (AB, BC, CD and AD).

No consumer is given direct connection from the feeders.

1.3 Why Higher Voltages for Transmission of Electric Power?

The transmission of electric power carried at higher voltages due to the following reasons,

(i) Reduces volume of conductor material:

Consider the transmission of electric power by a 3-phase line,

Power transmitted by the 3-phase line, $P = \sqrt{3}\ VI\ \cos\Phi$

$$\text{Load Current, } I = \frac{P}{\sqrt{3}\ VI\ \cos\Phi}$$

$$\text{Resistance/Conductor, } R = \frac{\rho l}{a}$$

Where ρ = resistivity of conductor material
a = area of cross section of conductor
l = length of the line

Total Power loss, $W = 3\,I^2R$

$$W = 3\left(\frac{P}{\sqrt{3}\,V\cos\Phi}\right)^2 \times \frac{\rho l}{a}$$

$$W = \frac{P^2\rho l}{V^2 \cos^2\Phi \cdot a}$$

Area of Cross-section, $a = \dfrac{P^2\rho l}{WV^2\cos^2\Phi}$

Total volume of the conductor material required $= 3\,a\,l$

$$= 3\left(\frac{P^2\rho l}{WV^2\cos^2\Phi}\right) \times l$$

$$= \frac{3P^2\rho l^2}{WV^2\cos^2\Phi} \rightarrow (1)$$

It is clear from equation (1) that for given values of P, l, ρ and W, the volume of conductor material required is inversely proportional to the square of transmission voltage. In other words, the greater the transmission voltage, the lesser is the conductor material required.

(ii) **Increases the power transfer:**

Power transmitted by the 3-phase line can be written as,

$$P = \frac{|V_S||V_R|}{X} \cdot \sin\delta$$

where P → Power transfer in MW/Phase
V_S → Sending end voltage in KV
V_R → Receiving end voltage in KV
δ → Phase angle between V_S and V_R
X_S → Sevics reactance of line

Maximum power transfer possible at $\delta = 90°$

Now, $$P_{max} = \frac{|V_S||V_R|}{X} \cdot \sin 90°$$

$$= \frac{|V_S||V_R|}{X}$$

By increasing rated voltage of transmission line (V_S and V_R), power limit also increases.

$$\therefore P \propto V^2$$

By increasing voltage by factor 2, power transmission decreases by factor 4.

(iii) **Increases transmission efficiency:**

Input power = output power + Total power

$$= P + \frac{P^2 \rho l}{V^2 \cdot \cos^2\Phi \cdot a}$$

Assuming 'J' to be the current density of the conductor, then $a = \frac{I}{J}$

$\therefore$ Input power $= P + \dfrac{P^2 \rho l}{V^2 \cdot \cos^2\Phi \cdot \left(\dfrac{I}{J}\right)}$

$$= P + \frac{P^2 \rho l J}{V^2 \cdot \cos^2\Phi \cdot I}$$

$$= P + \frac{P^2 \rho l J}{V^2 \cdot \cos^2\Phi} \times \frac{\sqrt{3}\, V \cos\Phi}{P} \quad \left[\because I = \frac{P}{\sqrt{3}\, V \cos\Phi}\right]$$

$$= P + \frac{\sqrt{3}\, P J \rho l}{V \cos\Phi}$$

Input power $= P\left[1 + \dfrac{\sqrt{3}\, J \rho l}{V \cos\Phi}\right]$

Transmission efficiency $= \dfrac{\text{Output power}}{\text{Input power}}$

$$= \frac{P}{P\left[1+\frac{\sqrt{3}\,J\rho l}{V\cos\Phi}\right]}$$

$$= \frac{1}{\left[1+\frac{\sqrt{3}\,J\rho l}{V\cos\Phi}\right]}$$

Transmission efficiency $=\left[1-\frac{\sqrt{3}\,J\rho l}{V\cos\Phi}\right]$ $\quad[\because \text{Binomial Theorem}]$

As J, ρ, l are constants, therefore, transmission efficiency increases when the line voltage is increased.

(iv) Decreases percentage line drop:

Line drop = I R

$$= I \times \frac{\rho l}{a}$$

$$= I \times \rho l \times \frac{J}{I} \qquad \left[\because a = \frac{I}{J}\right]$$

$$= \rho l J$$

% line drop $= \dfrac{J\rho l}{V} \times 100$

As J, ρ and l are constants, therefore percentage line drop decreases when the transmission voltage increases.

NOTE: It may appear advisable to use the highest positive voltage for transmission of electric power. But there is a limit to which this voltage can be increased. It is because increase in transmission voltage introduces insulation problem as well as the cost of switchgear, transformer and other terminal equipment is increased.

1.4 Conductors

Conductor is a physical medium to carry electrical energy from one place to other. It is an important component of overhead and underground electrical transmission and distribution systems. The choice of conductor depends on the cost and efficiency.

An ideal conductor should posses the following properties:

1. Maximum electrical conductivity to reduce copper losses and voltage drop.
2. High tensile strength so that it can withstand mechanical stresses.
3. Low specific gravity so that weight/unit volume is small.
4. Low installation and maintenance cost.

All the above properties are not found in a single material. Therefore while selecting a conductor material, compromise is made between the cost, electrical and mechanical properties and local conditions.

1.4.1 Conductor Materials

(a) **Copper:**

Copper is an ideal material for overhead lines. Copper conductors can be used as solid or stranded Copper conductors, Copper Conductor Steel Reinforced (C.C.S.R) and Cadmium Copper conductor alloy and its properties are,

1. The specific resistance 1.7 mΩ–cm is very low.
2. It has high tensile strength $\left(4 \times 10^6 \text{ N/m}^2\right)$
3. It can be soldered very easily.
4. It can be obtained in its purest form by electrolysis.
5. It has good weather proof resisting properties and long life.
6. It has higher scrap value.

Earlier year, the Copper was widely used for construction of overhead conductors but due to its non-availability within the country and high cost, it has been completely replaced by Aluminium material in the transmission system.

(b) **Aluminium:**

These became popularly used in transmission system due to its less cost and these are employed as solid or stranded Aluminium conductors.

The properties of these conductors are:

1. It is cheap as compared with the Copper.
2. The specific gravity of Aluminium is 2.7 gm/cm^3 and is much lower than that of Copper.
3. It is light in weight and is liable to greater swings.
4. For the same resistance and voltage drop, Aluminium material has 1.6 times the cross section of Copper but due to low density, the weight is only 0.48 times.
5. The conductivity of Aluminium is 60% that of Copper. Therefore for same resistance, the diameter of conductor has to be about 1.26 times the diameter of Copper conductor.
6. Its melting point is low and hence, under short circuits, the conductor may melt at joints.
7. It is not much affected even if used in chemical works.

Considering the contained properties of cost, conductivity, tensile strength, weight, availability etc., Aluminium is better than Copper. Therefore now-a-days Aluminium is being used as conductor material. It is particularly profitable to use Aluminium for heavy current transmission when the conductor size is large and it's cost forms a major proportion of the total cost of complete installation.

(c) **Aluminium and Steelreinforced (ASR):**

Conductors made of all Aluminium are not sufficiently mechanically strong for construction of long span lines. The deficiency in strength can be compensated by adding steel core to the Aluminium material. Such conductor is called as "Aluminium Conductor Steel Reinforced" (or) ACSR conductors.

In this composite material, steel provide the required mechanical strength to the Aluminium and it is having the following advantages.

1. The reinforcement with steel increases the tensile strength, but at the same time keeps the composite conductor light. Therefore,

A.S.R conductors will produce smaller sag and hence larger span can be used.
2. Towers of smaller heights can be used due to smaller sag with ASR.

(d) Galvanised steel:

Steel possesses high resistance (i.e., poor conductivity) as compared to other materials and is not used for transmission of electric power. Steel has a high tensile strength. Therefore, galvanised steel conductor can be used for extremely long spans.

Base steel conductors get rusted and hence never used. 'Galvanising' is essential to avoid rusting. Generally, galvanised steel wires are extensively used as 'earth wire' in distribution.

The properties of steel are as follows:

1. It has high tensile strength.
2. It's conductivity is the least.
3. It deteriorates rapidly due to rusting.

1.4.2 Types of Conductors

The following are the different types of conductors in transmission system:

1. Solid conductors
2. Stranded conductors
3. Composite stranded conductors

1. Solid conductors:

Solid conductors are typically smaller and stronger than stranded conductors. It consists of a single piece of metal wire and it is cheap for manufacturing. It is made up with solid Copper and its mechanical strength is high. Solid conductors are having the following disadvantages,

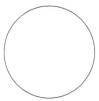

Fig. 1.3

1. It is difficult to transport.
2. Stringing is difficult because it cannot be bent easily.
3. Effective resistance increases.
4. Skin effect is high in A.C. System.

2. Stranded Conductors:

It is made up with Aluminium. Stranded conductors have two or more small cross section stands of conductor material twisted together to form a single conductor. Stranded conductors can carry the high currents and are usually more flexible than solid conductors.

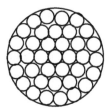

Fig. 1.4 Stranded Conductor

Standard conductors have the following advantages:

  i.   High electrical conductivity.
 ii.   Required Mechanical strength.
iii.   It can be bent easily.
 iv.   Stringing of the conductor is easy.
  v.   Transportation is easy because it is flexible.
 vi.   Skin effect is lesser than the solid conductor in A.C. system.

Stranded conductors are of the following types:

(a) All Aluminium Conductors (AAC):

This type is also called 'Aluminium stranded conductors'. It is made up of strands of electrical conductors grade Aluminium. AAC has conductivity about 61% IACS (International Annealed Copper Standard). Despite having a good conductivity, because of its relatively poor strength, AAC has limited use in transmission system and rural distribution system. However, AAC can be seen in urban areas for distribution where spans are usually short but higher conductivity is required.

(b) All Aluminium Alloy Conductors (AAAC):

These conductors are made from Aluminium alloy which is high strength Aluminium – Magnesium – Silicon alloy – this alloy conductor offers good electrical conductivity (about high strength Aluminium – Magnesium-Silicon alloy. This alloy conductor offers good electrical conductivity (about 52.5%IACS) with better mechanical strength. Because of AAAC's lighter weight as compared to ACSR of equal strength and current, et AAAC may be used for distribution purposes. It is not usually preferred for transmission. AAAC's can be employee in coastal areas because of their excellent corrosion resistance.

3. Composite stranded conductors:

In the stranded conductor, all the strands are Aluminium so the mechanical strength is not up to the mark. When the mechanical strength is less, sag will be increased and ground clearance to the conductor decreases. In order to increase the mechanical strength to the aluminium stranded conductor, steel strands will be added. Thesesteel strands can be placed at in inner layers of the conductor and aluminium strands are placed at outer layers. Cross section of this conductor can be utilised effectively and effective resistance will be decreased. Hence skin effect is further reduced to certain extent than the stranded conductors. Since this conductor is mix of steel and aluminium strand it is called as "Composite stranded conductors. These high strength conductors are normally used on long span distances and for minimum sag applications.

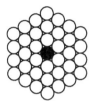

Fig. 1.5 Composite stranded conductors are of following types,

(a)   Aluminium Conductor Alloy Reinforced (ACAR):

ACAR conductor is formed by wrapping strands of high purity aluminium on high strength aluminium Magnesium – Silicon alloy core. ACAR has better electrical as well as mechanical properties them equivalent ACSR conductors. ACAR conductors may be used in overhead transmission, as well as distribution lines.

(b)   Aluminium Conductor Steel Reinforced (ACSR):

ACSR consists of a solid or stranded steel core with one or more layers of high purity aluminium strands wrapped in spiral. The core strands may be zinc coated (galvanised) steel or aluminium coated steel. Galvanization oraluminizationcoatings are thin and are applied to protect the steel from corrosion. The central steel core provides additional mechanical strength and hence sag is sufficiently less than other aluminium conductors. ACSR conductors are available in a wide range of steel content – from 6% to 40% ACSR with high steel content is selected where higher mechanical strength is required, such as river crossing. ACSR conductors are very widely used for all transmission and distribution purposes.

1.4.3 Construction of a Stranded Conductor

As mentioned above sections, stranded conductors are popularly used in transmission system due to their flexibility. In stranded conductors, there is generally one core strand and around this, successive layers of strands containing 6, 12, 18, 24, Strands. The standards of each layer are laid in helical fashion over the preceding layer is called stranding. Stranding is done in opposite direction to preceding layer. This mean, if the strands of one layer are twisted in clockwise direction the strands of next layer will be twisted in anticlockwise direction.

If 'x' is the no. of layers in the conductor, the total number of strands (N) in any conductor can be obtained as,

$$N = 3x^2 - 3x + 1$$

If the conductor having 3 layers, then

Fig. 1.6 Three Layer composite stranded Conductor

$$N = 27 - 9 + 1$$

$N = 19$ (fa 3 layer conductor consists 19 strands)

The Diameter of a conductor can be calculated as

$$D = (2x-1)d$$

Where, 'D' is the diameter of the conductor

'd' is the diameter each strand

Table representing the number of strands, Diameter and cross-sectional view of standard conductor for different number of layers:

SL NO	No. of layers 'x'	Total no. of strands $N = 3x^2 - 3x + 1$	Diameter of conductor $D = (2x-1)d$	Cross sectional view of stranded conductor
1	1	1	D	●
2	2	7	3d	

SL NO	No. of layers 'x'	Total no. of strands $N = 3x^2 - 3x + 1$	Diameter of conductor $D = (2x-1)d$	Cross sectional view of stranded conductor
3	3	19	5d	
4	4	37	7d	
5	5	61	9d	

1.5 Transmission Line Parameters

Transmission line consists of a set of conductors of proper size suitably spaced with respect to each other and insulated from the supporting poles or towers. The transmission line can be represented by an equivalent electrical circuit for the solution of it's performance under different conductors.

1. The conductors have definite 'Resistance' depending upon their dimensions and material.
2. The magnetic field produced by the current flowing through one conductor links with the other conductor causing "Inductance".
3. Between each conductor and the earth and also between the pairs of conductors, there is 'Capacitance'.
4. The insulation of the conductors may not be perfect and this result in a leakage of current to earth, the effect of which is represented by "conductance" assuming a leakage resistance is connected between the conductors and the earth.

Thus, the four parameters which represent a transmission line and describe its performance are,

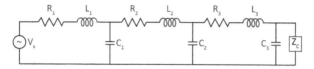

Fig. 1.7 Transmission Network

1. Resistance (R)
2. Inductance (L) and corresponding Inductive Reactance (X_L)
3. Capacitance (C) and corresponding capacitive reactance (X_C)
4. Conductance (G)

The values of these quantities have to be considered while analysing the performance of a transmission line. These parameters depend on the type of conductors used and their spacing with respect to each other, their clearance from ground etc. these are calculated per unit length of line and then for the entire length of the line. The line parameters are usually represented on 'loop – length basis' in case of single phase. 'Per conductor basis' in case of three – phase transmission.

These parameters and their calculations are discussed in the following sections.

1.5.1 Resistance

Though the contribution of line resistance to service line impedance can be neglected in most cases, it is the main source of line power loss. Thus, while considering transmission line economy the presence of line resistance must be considered.

The effective AC resistance is given by,

$$R = \frac{Average\, Power\, loss\, in\, conductor\, in\, watts}{I^2} \text{ Ohms}$$

Where 'I' is the current in amperes.
Ohmic DC resistance is given by,

$$R_O = \frac{\rho l}{A}$$

Where ρ = resistivity of the conductor ($\Omega - m$)
l = length of the conductor (m)
A = Cross sectional area (m²)

Resistance depends on following factors,

1. Spiralling
2. Temperature
3. Frequency
4. Current magnitude

1.5.2 Inductance

When an alternating current flows through a conductor, a changing flux is set up which links the conductor. Due to these flux linkages, the conductor possesses 'inductance'.

Mathematically, Inductance is defined as the flux linkages per ampere i.e.

Inductance, $L = \frac{\psi}{I}$ Henry

Where ψ – flux linkages in weber/turn

I – Current in amperes

1.5.3 Capacitance

Any two conductors separated by an insulating material constitute a capacitor. As any two conductors of an overhead transmission line are separated by air which acts as an insulation, therefore, capacitance exists between any two overhead line conductors.

Mathematically capacitance is defined as the charge per unit potential difference i.e.

Capacitance, $C = \frac{q}{V}$ farad

Where q – change on the line in coulomb

V – potential difference between the conductor in volts

When an alternating voltage is impressed on a transmission line, the charge on the conductors at any point increases anddecreases with the increase and decrease of the instantaneous value of the voltage between conductors at that point. The result is that a current flows between conductors is known as 'charging current'. This charging current flows in the line even when it is open circuited i.e., supplying no load. If affects the voltage drop along the line as well as the efficiency and power factors of the line.

1.5.4 Conductance

Conductance accounts for real power loss between conductors or between conductors and ground. For overhead lines, this power loss is due to leakage currents at insulators and to corona. Insulator leakage currents depends on the amount of dirt, salt and other contaminants that have accumulated on the surface, as well as on meteorological factors particularly the presence of moisture. Corona occurs when a high value of electric field strength at a conductor surface causes the air to become electrically ionized and to conduct. The real power loss due to corona called as 'corona loss' depends on meteorological conditions, particularly rain and conductor surface irregularities. Losses due to insulator leakage and corona are usually small compared to conductor I^2R less. Conductance is usually neglected in power system studies because it is very small component of the shunt admittance.

1.6 Skin Effect

Unequal distribution of current over the whole cross-section of the conductor is known as "skin effect".

When a conductor is carrying steady Direct Current (D.C), this current is uniformly distributed over the whole cross-section of the conductor. However, an Alternating Current (A.C) flowing through the conductor does not distribute uniformly, but it has a tendency to concentrate near the skin i.e., surface of conductor as shown in fig.1.8.

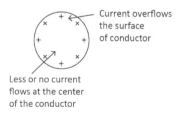

Fig. 1.8 Solid Conductor

The cause of skin effect can be easily explained as follows,

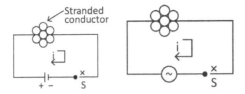

Fig. 1.9 (a) Skin effect in D.C. Circuit   Fig. (b) Skin effect in A.C. Circuit

A stranded conductor is connected to the D.C. supply as shown in Fig.1.9 (a). When the switch 'S' is closed, current start to flow in the circuit and conductor. Since supply is D.C., current will distribute over the surface of the conductor uniformly. Same conductor connected to A.C. supply as shown in Fig. (b). When the supply is given by closing the switch, current in the conductor will not distribute uniformly because inductance offered by the outer strands will be different than the inner strand. As this conductor consists of 6 outer strands and 1 inner strand, flux produced by the outer stands should link with inner strand and links with it. This flux should not link with air because of high reluctance. Now flux produced by inner strand links with itself and also having the flux linkages by outer strand. So total flux linkage of inner strand is more than the outer strand.

Now as we know that,

$$L = \frac{N\phi}{I}$$

Where, N is no. of turn
ϕ is flux
I is current

From above equation, inductance offered by the inner strand is more as flux linkages is more, then the currentflowing in the inner strand is less. Inductance offered by the outer strand is less as flux linkages are less, so more current flowing over the surface. This unequal distribution of current over the surface of the conductor is called skin effect.

Skin effect depends on the following factors:

1. Nature of material
2. Frequency: The skin effect increases with the increase in frequency.
3. Diameter of conductor: The skin effect increases with the square of the diameter of the conductor.
4. Type of conductor: The skin effect is less for stranded conductor than the solid conductor.

1.7 Proximity Effect

Like skin effect, 'Proximity Effect' also results in non-uniform distribution of current in cross-section of a conductor due to the presence of other current carrying conductor in its proximity (i.e., vicinity)

Conductor A   Conductor B

Fig. 1.10

Suppose there is two conductors 'A' and 'B' placed near to each other as shown above Fig. when conductor 'A' carries a current, it's magnetic flux links with the nearby conductor 'B'. However, more flux links with the nearer had of the conductor 'B' (shown shaded in Fig) there with further both (shown non-shaded in Figure) If both conductors carry currents, then both conductors produce magnetic flux linking with itself and with another conductor and hence resultant magnetic flux has to be considered. If the conductor carry currents in 'opposite directions', then the magnetic fields set up will cause an increase in the resultant magnetic flux linkage and hence also in the current density in the adjacent portions of the conductor. On the other hand, if the currents are in the 'same direction', then there is increase in the resultant magnetic flux linkage and hence in the current density in remote parts of the conductors.

'Proximity effect' affects the current distribution and results in an increase in the effective resistance of the conductors, decrease of self inductance and increases power loss in the conductors.

1.8 Concept of GMR & GMD

Transmission line is basically a 3ϕ, 3 wire system and it can be represented in an electrical network model in terms of resistance, inductance and capacitance. In order to calculate inductance and capacitance in a phase manner at a faster rate, the mathematical concepts are proposed and they are GMR & GMD.

If the system is dealing with space coordinates, then we can approach 'Geometric Mean Analysis'.

If the system is dealing with plane coordinates, then we can approach 'Arithmetic Analysis'.

Transmission network is dealing with the space coordinates; we can adopt 'Geometric Mean Analysis'.

Geometric Mean Radius (GMR) (or) Self GMD:

It is a distance of a point from the centre of the conductor to towards the circumference with respect to same point in the space.

$GMR = r^l$

$r^l \rightarrow$ Imaginary radius. Due to skin effect, we can't take actual radius as it is but imaginary radius. (i.e. r^l)

Geometric Mean Distance (or) Mutual GMD:

Consider a point 'K' in the space is surrounded by 'n' other points. The equivalent distance of point 'K' w.r.t. to 'n' other points will be the geometric mean of the individual distances between the reference point and each individual point.

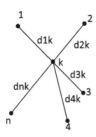

Fig. 1.11 GMD of a conductor

$$\therefore GMD_K = \sqrt[n]{d_{1K} \cdot d_{2K} \cdot d_{3K} \cdot d_{4K} \ldots \ldots d_{nK}}$$

Physically Transmission lines are of two configurations:

1. Symmetrical configuration (or) Equilateral spacing configuration
2. Asymmetrical (or) Flat (or) Horizontal configuration.

1. Symmetrical configuration:

GMR of conductor R, $GMR_R = r^l$
GMR of conductor Y, $GMR_Y = r^l$
GMR of conductor B, $GMR_B = r^l$

$$GMR = \sqrt[3]{GMR_R \cdot GMR_Y \cdot GMR_B}$$

$$= \sqrt[3]{r^l \cdot r^l \cdot r^l}$$

$$= \sqrt[3]{(r^l)^3}$$

$GMR = r^l = GMR_R = GMR_Y = GMR_B$
$GMR = 0.7788\ r$

GMD:

$$GMD_R = \sqrt{GMD_Y \cdot GMD_B} = \sqrt{d \cdot d} = d$$

$$GMD_Y = \sqrt{GMD_B \cdot GMD_R} = \sqrt{d \cdot d} = d$$

$$GMD_B = \sqrt{GMD_R \cdot GMD_Y} = \sqrt{d \cdot d} = d$$

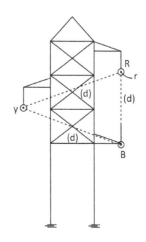

∴ GMD of the configuration, $GMD = \sqrt[3]{GMD_R \cdot GMD_Y \cdot GMD_B}$

$$= \sqrt[3]{d \cdot d \cdot d}$$

$GMD = d$ for equilateral spacing configuration.

2. **Asymmetrical configuration:**

GMR of 'R' conductor, $GMR_R = r^l$
GMR of 'Y' conductor, $GMR_Y = r^l$
GMR of 'B' conductor, $GMR_B = r^l$
GMR of the configuration
$GMR = GMR_R = GMR_Y = GMR_B = 0.7788\, r$

GMD:

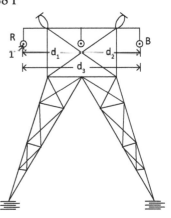

$GMD_R = \sqrt{d_1 \times d_3}$

$GMD_Y = \sqrt{d_1 \times d_2}$

$GMD_B = \sqrt{d_2 \times d_3}$

GMD of the configuration,

$GMD = \sqrt[3]{GMD_R \cdot GMD_Y \cdot GMD_B}$

$= \sqrt[3]{\sqrt{d_1 d_3}\sqrt{d_1 d_3}\sqrt{d_2 d_3}}$

$= \sqrt[3]{\sqrt{d_1^2 d_2^2 d_3^2}}$

$= \sqrt[3]{d_1 d_2 d_3}$

∴ $GMD = \sqrt[3]{d_1 d_2 d_3}$ (or) if it is x, y, z then $GMD = \sqrt[3]{xyz}$

NOTE:
1. GMR depends only on radius of the conductor and it is independent of the line to line space (distance).
2. GMD of the conductor depends on the configuration but not on the radius.

1. What is the self GMD of the given configuration?

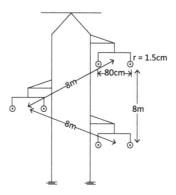

Self GMD = Self GMD_R

Self $GMD_R = \sqrt{Self\ GMD_{R_1} \cdot Self\ GMD_{R_2}}$

Self GMD = Self GMD_{R_1}

since sub conductors are symmetric

Self $GMD_{R_1} = \sqrt{r^1 \times s}$

$= \sqrt{0.7788 \times r \times s}$

$= \sqrt{0.7788 \times 1.5 \times 80} = 9.66$ cm

Self GMD_{R_1} = Self GMD_{R_2} = Self GMD_R

Now, GMD = d = 8m   (∵ Symmetric configuration)

NOTE: In case of bundled conductors, consider the sub conductors effect for calculation of self GMD and ignore the sub conductor effect for calculation of GMD.

2. Find the GMR & GMD of given configuration.

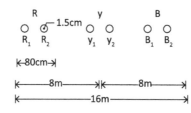

$$\text{Self GMD} = \text{Self GMD}_{R_1} = \sqrt{0.7788 \times 1.5 \times 80}$$

$$= 9.66 \text{ cm}$$

$$\text{GMD} = \sqrt[3]{xyz}$$

$$= \sqrt[3]{5 \times 8 \times 16}$$

$$= 4.5 \text{ m}$$

1.9 Line Inductance

Inductance is the property by virtue of which a circuit oppose changes in the value of a varying current flowing through it. While the resistance of a circuit opposes flow of both steady as well as varying currents. The inductance causes opposition only to varying currents. Inductance does not cause any opposition to steady currents. In case of transmission & distribution lines the current flowing is alternating, the effect of inductance is therefore to be considered the opposition to the flow of varying current owing to inductance is viewed as voltage drop.

It is well known fact that a current carrying conductor is surrounded by concentric circles of magnetic lines. In case of AC system this field set up around the conductor i.e. not constant but changing and links with the same conductor as well as with other conductors. Due to these flux linkages, the line posse's inductances. Thus for determination of inductance of a circuit, determination of flux linkages is essential.

Flux linkages of a conductor:

A log straight cylindrical conductor carrying a current is surrounded by a magnetic field. The magnetic lines of force will exist inside the conductor as well as outside the conductor. Both of these fluxes contribute to the inductance of a conductor.

1.9.1 Flux Linkages of a Conductor Due to Internal Flux

Consider a long straight cylindrical conductor of radius 'r' meters and carrying a current of 'I' amp as shown In fig.

Lets us denote magnetic field intensity at any point distant 'x' meters from the centre of the conductor by 'H_x'.

Ampere Circuital Law: Line integral of magnetic field intensity 'H' around a closed path is equal to the current enclosed by that path.

$$\oint_L H \cdot dl = I$$

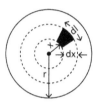

Fig. 1.12 Flux Linkages due to internal Flux

From the Ampere's circuital law,

m.m.f around any closed path is equal to the current enclosed by that path.

$$\oint H_x \cdot dl = I_x$$

$$H_x \cdot \oint dl = I_x$$

$$H_x \cdot 2\pi k = I_x$$

The current inside a lines of force of distance 'x' is

$$I_x = I \cdot \frac{\pi x^2}{\pi r^2}$$

$$I_x = I \cdot \frac{x^2}{r^2}$$

Field intensity inside the conductor at a distance 'x' from the centre,

$$H_x = \frac{I_x}{2\pi x}$$

$$= \frac{1}{2\pi x} \cdot I \cdot \frac{x^2}{r^2}$$

$$H_x = I \cdot \frac{x}{2\pi r^2} \text{ AT/m}$$

Flux density $B_x = \mu H_x$ wb/m²

$B_x = \mu_0 H_x$ $[\because \mu = \mu_0 \mu_r, \mu_r = 1$ for non magnetic material]

$$B_x = \mu_0 \cdot \frac{I}{2\pi r^2} \text{ wb}/m^2$$

Flux enclosed in element of thickness 'dx' and per meter length of conductor is given by

$$d\phi = B_x \times l \times dx$$

$$d\phi = B_x \times l \times dx$$

$$d\phi = \mu_0 \cdot \frac{Ix}{2\pi r^2} dx \text{ webers/m}$$

But this flux links with only the current lying within the circle of distance 'x'. So, flux linkages per meter length of the conductor is,

$$d\psi = \frac{\pi x^2}{\pi r^2} \cdot d\phi$$

$$= \frac{x^2}{r^2} \cdot \frac{\mu_0 Ix}{2\pi r^2} \cdot dx$$

$$d\psi = \mu_0 \cdot \frac{Ix^3}{2\pi r^4} \cdot dx \text{ wb-turn/m}$$

∴ Total flux linkages from centre of conductor to the surface of the conductors is,

$$\psi_{int} = \int_0^r \frac{\mu_0 Ix^3}{2\pi r^4} dx$$

$$= \frac{\mu_0 I}{2\pi r^4} \int_0^r x^3 dx$$

$$= \frac{\mu_0 I}{2\pi r^4} \frac{(x^4)}{4} \Big|_0^r$$

$$\psi_{int} = \frac{\mu_0 I}{2\pi r^4} \cdot \frac{r^4}{4}$$

$$\psi_{int} = \frac{\mu_0 I}{8\pi} \Rightarrow \frac{4\pi \times 10^{-7} \cdot I}{8\pi}$$

$$\psi_{int} = \frac{1}{2} \times 10^{-7} \cdot I \text{ wb-turn/m} \quad \rightarrow (1)$$

1.9.2 Flux Linkages of a Conductor Due to External Flux

Consider two points 1 & 2 distant d_1 and d_2 from the centre of the conductor. Since the flux paths are concentric circles around the conductor, whole of the flux between points 1 and 2 lies within the concentric cylindrical surfaces passing through this points 1 and 2.

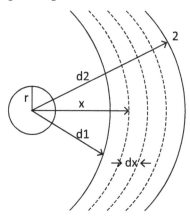

Fig. 1.13 Flux linkages due to external Flux

The field intensity at any distance 'x' from the centre of the conductor $(x > r)$,

$$H_x = \frac{I}{2\pi x} \text{ AT/m}$$

Flux density $B_x = \mu_0 H_x = \dfrac{\mu_0 I}{2\pi x}$ wb/m²

So the flux through the cylindrical shell of radial thickness 'dx' and axial length one meter.

$$d\phi = B_x \times 1 \times dx$$

$$d\phi = B_x dx$$

$$d\phi = \frac{\mu_0 I}{2\pi x} dx \text{ wb/m}$$

Now, flux linkages per meter is equal to $d\phi$, since flux external to conductor links all the current in the conductor once and only once.

$$\therefore d_\psi = d\phi = \frac{\mu_0 I}{2\pi x} dx \text{ wb-turn/m}$$

Total, flux linkages between points 1 & 2,

$$\psi_{ext} = \int_{d_1}^{d_2} \frac{\mu_0 I}{2\pi x} dx$$

$$= \frac{\mu_0 I}{2\pi} \int_{d_1}^{d_2} \frac{1}{x} dx$$

$$= \frac{\mu_0 I}{2\pi} \left(\log_e x\right)_{d_1}^{d_2}$$

$$= \frac{4\pi \times 10^{-7}}{2\pi} \cdot I \log_e \frac{d_2}{d_1}$$

$$\psi_{ext} = 2 \times 10^{-7} \cdot I \cdot \log_e \frac{d_2}{d_1} \text{ wb-turn/m} \rightarrow (2)$$

1.9.3 Inductance of 1 – φ Two Wire Line

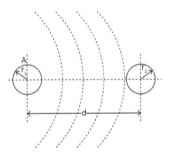

Fig. 1.14

Consider a 1-φ line consisting of two parallel conductors A and B of radius r_1 and r_2 spaced 'd' meters apart. Conductor A and B carry the same current in magnitude but opposite in directions as one forms the return path for the other. The inductance of each conductor is due to internal flux linkages and external flux linkages.

It can be assumed that all the flux produced by current in conductor A links all the currents up to the centre of the conductor B and that the flux beyond centre of the conductor B does not link any current.

Based on the above assumption, flux linkages of conductor 'A' due to external flux can be obtained from equation (2) by substituting $d_2 = d$ and $d_1 = r_1$. Thus flux linkages of conductor A due to external flux only.

$$\psi_{Aext} = 2\times 10^{-7} . I . \log_e \frac{d}{r_1} \text{ wb-turn/m}$$

Flux linkages of conductors A due to internal flux only

$$\psi_{int} = \frac{1}{2}\times 10^{-7} . I \text{ wb-turn/m}$$

Total flux linkages of conductor A

$$\psi_A = \psi_{Aext} + \psi_{Aint}$$

$$= \left[2\times 10^{-7} I \log_e \frac{d}{r_1} + \frac{1}{2} I \times 10^{-7}\right]$$

$$= 2\times 10^{-7} . I \left[\log_e \frac{d}{r_1} + \frac{1}{4}\right]$$

$$= 2\times 10^{-7} \left[\log_e \frac{d}{r_1} + \log_e e^{\frac{1}{4}}\right]$$

$$= 2\times 10^{-7} . I \left[\log_e \frac{d}{(r_1)^1}\right]$$

Where $r_1^1 \quad r_1 e^{-\frac{1}{4}}$

$$\psi_A = 2\times 10^{-7} . I . \log_e \frac{d}{(r_1)^1}$$

Inductance of conductor A, $L_A = \dfrac{\psi_A}{I}$

$$\therefore L_A = 2\times 10^{-7} \log_e \frac{d}{(r_1)^1} \text{ H/m}$$

The product $\left(r_1 . e^{-\frac{1}{4}}\right)$ is known as GMR of the conductor and is equal to 0.7788 times of the conductors.

Similarly, Inductance of conductor B,

$$L_B = 2\times 10^{-7} \log_e \frac{d}{(r_2)^1} \text{ H/m}$$

∴ The inductance of the line, $L = L_A + L_B$

If $r_1^1 = r_2^1 = r$, $L = 2\times 10^{-7} \left(\log_e \dfrac{d}{r^1} + \log_e \dfrac{d}{r^1}\right)$

$$\therefore L = 4 \times 10^{-7} \log_e \frac{d}{r^1} \text{ H/m}$$

(or)

$$L = 0.4 \times ln \frac{d}{r^1} \text{ mH/km}$$

1.10 Flux Linkages of One Conductor in a Group of Conductors

Consider a group of parallel conductors 1, 2, 3, N carrying current $I_1, I_2, I_3, \ldots I_n$ respectively. Let it be assumed that the sum of the currents in various conductors is zero. i.e.

$$I_1 + I_2 + I_3 + \ldots + I_n = 0$$

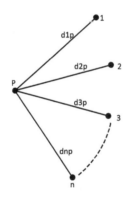

Fig. 1.15 Flux linkages of a conductor in a group

Theoretically, the flux due to a conductor extends from the centre of the conductor right up to infinity but let us assume that the flux lineages extend up to remote point 'P' and the respective distances are as marked in above.

The current in each conductor sets up a certain flux due to its own current. The sum of all these fluxes is the total flux of the system and total flux linkages of any one conductor is the sum of its linkages with all the individual fluxes set up by the conductors of the system.

Now let us determine the flux linkages of conductor 1 due to current I_1 carried by the conductor itself and the flux linkages of conductor 1 due to current carried by other Conductors (2, 3, n)

The flux lineages of conductor 1 due to its own current I_1 (Internal & external) up to point P,

$$\psi_{1P_1} = 2\times 10^{-7}. I_1 .ln\frac{d_{1p}}{r_1^1} \text{ wb - turn}$$

The flux linkages of conductor 1 due to current in conductor 2

$$\psi_{1P_2} = 2\times 10^{-7}. I_2 .ln\frac{d_{2p}}{d_{12}}$$

Thus the expression for flux linkages of conductor 1 due to currents in all conductors can be written as,

$$\psi_{1P} = 2\times 10^{-7}\left[I_1 ln\frac{d_{1P}}{r_1^1} + I_2 ln\frac{d_{2P}}{d_{12}} + I_3 ln\frac{d_{3P}}{d_{13}} + ... + I_n ln\frac{d_{nP}}{d_{1n}} \right]$$

The above equation may be written as,

$$\psi_{1P} = 2\times 10^{-7}\left[I_1 ln\frac{1}{r_1^1} + I_2 ln\frac{1}{d_{12}} + I_3 ln\frac{1}{d_{13}} + ... + I_n ln\frac{1}{d_{1n}} \right]$$
$$+ 2\times 10^{-7}\left[I_1 ln d_{1P} + I_2 ln d_{2P} + I_3 ln d_{3P} + ... + I_n ln d_{nP} \right]$$
$$\rightarrow (3)$$

To account for the total, flux linkages to conductor 1, the point 'P' must approach infinity and in this condition,

$$d_{1P} \cong d_{2P} \cong d_{3P} \cong ... d_{np} \cong d$$

then $\lim_{d\to\infty}(I_1 + I_2 + I_3 + ... + I\_n)\ ln d = 0$

$$[\because I_1 + I_2 + I_3 + ... + I_n = 0]$$

Equation (3) simplifies and the equation for the flux linkages to conductor 1 becomes

$$\psi_1 = 2\times 10^{-7}\left[I_1 ln\frac{1}{r_1^1} + I_2 ln\frac{1}{d_{12}} + I_3 ln\frac{1}{d_{13}} + ... + I_n ln\frac{1}{d_{1n}} \right] \text{ wb-turn} \quad \rightarrow (4)$$

1.11 Inductance of 3φ Single Circuit Overhead Transmission Law

(a) With Equal Spacing (or) Symmetrical spacing:

Consider 3φ line with conductors A, B and C each of this is 'r' radius and these are spaced at the corners of an equilateral triangle each side being 'd'.

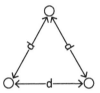

Fig. 1.16 Equilaterally Spaced Conductors

The currents in conductors are I_A, I_B and I_C respectively which satisfy the relationship $I_A + I_B + I_C = 0$.

Flux linkages of conductor 'A' due to itself and current flowing through other conductors.

$$\psi_A = 2\times 10^{-7}\left[I_A \ln\frac{1}{r^1}+I_B \ln\frac{1}{d}+I_C \ln\frac{1}{d}\right]$$

$$= 2\times 10^{-7}\left[I_A \ln\frac{1}{r^1}+(I_B+I_C)\ln\frac{1}{d}\right]$$

$$= 2\times 10^{-7}\left[I_A \ln\frac{1}{r^1}-I_A \ln\frac{1}{d}\right] \qquad [\because I_B+I_C=-I_A]$$

$$\psi_A = 2\times 10^{-7}\left[I_A \ln\frac{1}{r^1}-I_A \ln\frac{1}{d}\right]$$

$$\psi_A = 2\times 10^{-7} I_A \left[\ln\frac{d}{r^1}\right] \text{ wb-turn/m}$$

Inductance of conductor 'A', $L_A = \dfrac{\psi_A}{I_A} \Rightarrow 2\times 10^{-7}\left[\ln\dfrac{d}{r^1}\right]$

$$\therefore L_A = 0.2\ln\frac{d}{r^1} \text{ mH/Km} \qquad \rightarrow(5)$$

(b) With unsymmetrical spacing:

Consider a 3ϕ line with conductors A, B and C each of radius 'r' meter. Let the spacing between them be d_1, d_2 and d_3 the current flowing through them be I_A, I_B and I_C respectively.

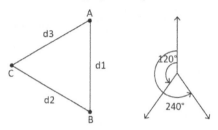

Fig. 1.17 Unsymmetrical spaced Conductors

The flux linkages of conductor 'A' due to its own current I_A and other conductor currents I_B and I_C.

$$\psi_A = 2\times10^{-7}\left[I_A \ln\frac{1}{r^1}+I_B \ln\frac{1}{d_1}+I_C \ln\frac{1}{d_3}\right] \rightarrow (6)$$

$$\psi_B = 2\times10^{-7}\left[I_B \ln\frac{1}{r^1}+I_A \ln\frac{1}{d_1}+I_C \ln\frac{1}{d_2}\right] \rightarrow (7)$$

$$\psi_C = 2\times10^{-7}\left[I_C \ln\frac{1}{r^1}+I_A \ln\frac{1}{d_3}+I_B \ln\frac{1}{d_1}\right] \rightarrow (8)$$

If the system is balanced,

$$I_A = I_B = I_C = I$$

Taking I_A as reference phases, the current are represented in symbolic form as,

$$I_A = I \angle 0°, \quad I_B = I \angle 240°, \quad I_C = I \angle 120°$$

$$I_B = I(\cos 240° + j\sin 240°) = I(-0.5 - j\, 0.866)$$

$$I_C = I(\cos 120° + j\sin 120°) = I(-0.5 + j\, 0.866)$$

Substituting value of I_B and I_C in equation (6),

$$\psi_A = 2\times10^{-7}\left[I\ln\frac{1}{r^1}+I(-0.5-j0.866)\ln\frac{1}{d_1}+I(-0.5+j0.866)\ln\frac{1}{d_3}\right]$$

$$\psi_A = 2\times 10^{-7}.I\left[\ln\frac{1}{r^1}-0.5\ln\frac{1}{d_1}-0.5\ln\frac{1}{d_3}-j0.866\ln\frac{1}{d_1}+j0.866\ln\frac{1}{d_3}\right]$$

$$= 2\times 10^{-7}.I\left[\ln\frac{1}{r^1}+\frac{1}{2}\left(\ln\frac{1}{d_1 d_3}\right)+j\frac{\sqrt{3}}{2}\left(\ln\frac{1}{d_3}-\ln\frac{1}{d_1}\right)\right]$$

$$= 2\times 10^{-7}.I\left[\ln\frac{1}{r^1}+\frac{1}{2}\ln d_1 d_3 + j\frac{\sqrt{3}}{2}.\ln\frac{d_1}{d_3}\right]$$

$$\psi_A = 2\times 10^{-7}.I\left[\ln\frac{1}{r^1}+\frac{1}{2}\sqrt{d_1 d_3}+j\sqrt{3}\ln\sqrt{\frac{d_1}{d_3}}\right]\text{ wb-turn/m}$$

$$\psi_B = 2\times 10^{-7}.I\left[\ln\frac{1}{r^1}+\sqrt{d_1 d_2}+j\sqrt{3}\ln\sqrt{\frac{d_2}{d_1}}\right]\text{ wb-turn/m}$$

$$\psi_C = 2\times 10^{-7}.I\left[\ln\frac{1}{r^1}+\sqrt{d_2 d_3}+j\sqrt{3}\ln\sqrt{\frac{d_3}{d_2}}\right]\text{ wb-turn/m}$$

Inductance of conductor A, $L_A = \dfrac{\psi_A}{\psi_B} = \dfrac{\psi_A}{I}$

$$\therefore L_A = 2\times 10^{-7}\left[\ln\frac{1}{r^1}+\ln\sqrt{d_1 d_3}+j\sqrt{3}\ln\sqrt{\frac{d_1}{d_3}}\right]\text{ H/m}$$

$$L_B = 2\times 10^{-7}\left[\ln\frac{1}{r^1}+\ln\sqrt{d_1 d_2}+j\sqrt{3}\ln\sqrt{\frac{d_2}{d_1}}\right]\text{ H/m}$$

$$L_C = 2\times 10^{-7}\left[\ln\frac{1}{r^1}+\ln\sqrt{d_2 d_3}+j\sqrt{3}\ln\sqrt{\frac{d_3}{d_2}}\right]\text{ H/m}$$

Thus we see that when the conductors of a 3ϕ transmission lines are not equidistant from each other i.e., unsymmetrically spaced, the flux linkages and inductances of various phases are different which cause unequal voltage drops in three phases and transfer of power between phases due to mutual inductances even if the currents in the conductors are balanced. The unbalancing effectof account of irregular spacing of conductors is avoided by transposition of conductors.

Transposition of Transmission lines:

Changing the position of power conductors at regular intervals by maintaining equal distance so that the position of the original phase conductor will be replaced by its successive phase conductor.

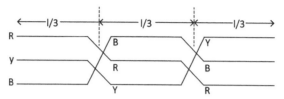

Fig. 1.18 Transposition of conductors

With the transposition, making resultant flux as zero by which emf induced on the telecommunication lines becomes zero.

(c) **Unsymmetrical spacing with transposition:**

The effect of transposition is that each conductor has the same average inductance which is given as

$$L = \frac{L_A + L_B + L_C}{3}$$

$$= \frac{1}{3} \cdot 2 \times 10^{-7} \left[3\ln\frac{1}{r^1} + \ln\sqrt{d_1 d_3} + \ln\sqrt{d_1 d_2} + \ln\sqrt{d_2 d_3} \right.$$

$$\left. + j\sqrt{3}(\ln\sqrt{\frac{d_1}{d_3}} + \ln\sqrt{\frac{d_2}{d_1}} + \ln\sqrt{\frac{d_3}{d_2}}) \right]$$

$$= 2\times 10^{-7} \left[\ln\frac{1}{r^1} + \frac{1}{3}\ln d_1 d_2 d_3 + j\frac{\sqrt{3}}{3}\ln 1 \right]$$

$$L = 2\times 10^{-7} \left[\ln\frac{1}{r^1} + \ln\sqrt[3]{d_1 d_2 d_3} \right]$$

$$L = 2\times 10^{-7} \left[\ln\frac{1}{r^1} + \ln\sqrt[3]{d_1 d_2 d_3} \right]$$

$$L = 2\times 10^{-7} \ln\frac{\sqrt[3]{d_1 d_2 d_3}}{r^1} \text{ H/m}$$

(i) **When the conductors of 3φ tr. Line are in the same plane: (Horizontal Plane)**

Horizontal plane arrangement is shown in Fig.
$$d_1 = d, \quad d_2 = d, \quad d_3 = 2d$$

```
A         B         C
O         O         O
|←—d—→|←—d—→|
|←———2d———→|
```

Inductance of conductor A,

$$L_A = 2\times 10^{-7}\left[\ln\frac{1}{r^1} + \ln\sqrt{2d\times d} + j\sqrt{3}\ln\frac{d}{2d}\right]$$

$$= 2\times 10^{-7}\left[\ln\frac{1}{r^1} + \frac{1}{2}\ln 2d^2 + j\frac{\sqrt{3}}{2}\ln\frac{1}{2}\right]$$

$$= 2\times 10^{-7}\left[\ln\frac{1}{r^1} + \frac{1}{2}\ln 2 + \frac{1}{2}\ln d^2 - j\frac{\sqrt{3}}{2}\ln 2\right]$$

$$= 2\times 10^{-7}\left[\ln\frac{1}{r^1} + \frac{1}{2}\ln 2 + \ln d - j\,0.866\ln 2\right]$$

$$L_A = 2\times 10^{-7}\left[\ln\frac{d}{r^1} + \frac{1}{2}\ln 2 + \ln d - j\,0.866\ln 2\right]\; \text{H/m}$$

$$L_B = 2\times 10^{-7}\left[\ln\frac{1}{r^1} + \ln\sqrt{d\times d} + j\ln\frac{d}{d}\right]$$

$$L_B = 2\times 10^{-7}\ln\frac{d}{r^1}\; \text{H/m}$$

$$L_C = 2\times 10^{-7}\ln\frac{d}{r^1} + \frac{1}{2}\ln 2 + j\,0.866\ln 2\; \text{H/m}$$

(ii) **When the conductors are at corners of right angled triangle:**
Substituting $d_1 = d_2 = d$ and $d_3 = \sqrt{2}d$

$$L_A = 2 \times 10^{-7} \left[\ln \frac{1}{r^1} + \ln \sqrt{\sqrt{2} d \times d} + j\sqrt{3} \ln \sqrt{\frac{d}{\sqrt{2}d}} \right]$$

$$= 2 \times 10^{-7} \left[\ln \frac{1}{r^1} + \frac{1}{2} \ln \sqrt{2} d^2 + j\sqrt{3} \ln \sqrt{\frac{1}{\sqrt{2}}} \right]$$

$$L_A = 2 \times 10^{-7} \left[\ln \frac{d}{r^1} + \frac{1}{2} \ln \sqrt{2} - j\frac{\sqrt{3}}{2} \ln \sqrt{2} \right] H/m$$

$$L_B = 2 \times 10^{-7} \left[\ln \frac{1}{r^1} + \ln \sqrt{d \times d} + \sqrt{3} \ln \sqrt{\frac{d}{d}} \right]$$

$$L_B = 2 \times 10^{-7} \left[\ln \frac{1}{r^1} + \ln d \right] H/m$$

$$L_C = 2 \times 10^{-7} \left[\ln \frac{d}{r^1} + \frac{1}{2} \ln \sqrt{2} + j\frac{\sqrt{3}}{2} \ln \sqrt{2} \right] H/m$$

Problems

1. A 3ϕ, 33KV, OHTL, 30 km long has its conductor ACCR 15mm diameter spaced at the corners of an equilateral triangle of 1.5m side. Find the 2/phase of the system.

Sol: Conductor radius $r = \frac{15}{2} mm = 7.5\ mm$

GMR of the conductor $r^1 = 0.7788 \times r$

$= 0.7788 \times 7.5$

$= 5.841 mm$

Spacing of conductors $d = 1.5m = 1500mm$

Inductance per phase $L = 2 \times 10^{-7} \ln \frac{d}{r^1}$

$= 2 \times 10^{(-7)} \ln \frac{1500}{7.5}$

$= 1.11\ mH/km$

Electrical Design of Overhead Transmission Lines | 51

2. **Find the inductance/km of a 3φ tr. Line using 20mm diameter conductors when conductors are situated at the corners of a triangle with spacing of 4, 5 and 6 m. conductors are regularly transposed.**

Sol: Conductor radius $r = \dfrac{20}{2} = 10mm = 1.0\ cm$

GMR of the conductor $r' = 0.7788 \times 1 = 0.7788\ cm$

Inductance per phase $L = 0.2\ln \dfrac{\sqrt[3]{d_1 d_2 d_3}}{r'}$

$= 0.2\ln \dfrac{\sqrt[3]{400 \times 500 \times 600}}{0.7788}$

$= 1.29\ mA/km$

3. **The three conductors of a 3φ tr. line are arranged in a horizontal plane and are 3m apart. The diameter of each conductor is 4 cm. find the inductance per km of each phase. Assume balanced Load & RYB phase sequence.**

Sol: Diameter of each conductor, $D = 4cm$

Radius of each conductor, $R = \dfrac{4}{2} = 2cm$

GMR of the conductor, $r' = 0.7788 \times 2 = 1.5576 cm$

Space between the conductors, $d = 3m = 300cm$

Inductance of conductor A, $L_A = 0.2\left[\ln\dfrac{d}{r'} + \dfrac{1}{2}\ln 2 - j\ 0.866 \ln 2\right]$

$L_A = 0.2\left[\ln\dfrac{300}{1.537} + \dfrac{1}{2}\ln 2 - j\ 0.866 \ln 2\right]$

$L_A = 0.2\left[5.26 + 0.346 - j0.866 \times 0.6931\right]$

$L_A = (1.12 - j0.12)\ mH/km$

$L_B = 0.2\ln\dfrac{d}{r'}$

$= 0.2\ln\left(\dfrac{300}{1.537}\right) = 1.052\ MH/Km$

$$L_C = 0.2 \left(\ln\frac{d}{r^1} + \frac{1}{2}\ln 2 + j\,0.866\ln 2 \right)$$

$$L_C = (1.12 + j0.12)\,\text{MH/Km}$$

4. A 3φ OHTL is designed with an equilateral spacing of 2.5m with a conductor diameter of 1.2cm. if the line is constructed with horizontal spacing with suitably transposed conductors, find the spacing between adjacent conductors which would give the same value of inductance as in the equilateral arrangement.

Sol: Spacing of conductors $d = 3.5m = 350mm$

Conductor diameter $D = 1.2cm$, Radius, $r = \dfrac{1.2}{2} = 0.6cm = 6mm$

GMR of the conductor, $r^1 = 0.7788 \times 6 = 4.67mm$

For equilateral spacing arrangement,

Inductance / phase $L = 0.2\ln\dfrac{d}{r^1}$ mH/Km

$$= 0.2\ln\left(\frac{3500}{4.67}\right)$$

$$= 1.324 \times 10^{-6} \text{ H/m}$$

If the same system is arranged horizontally, and it is transposed, then inductance of a transposed system,

$$L = 0.2\ln\left[\frac{\sqrt[3]{d_1 d_2 d_3}}{r^1}\right] \text{mH/Km}$$

$$1.32 \times 10^{-6} = 2 \times 10^{-7} \ln\frac{\sqrt[3]{d_1 d_2 d_3}}{r^1} \text{ H/m}$$

$$\ln\frac{\sqrt[3]{d_1 d_2 d_3}}{r^1} = \frac{1.32 \times 10^{-6}}{2 \times 10^{-7}}$$

$$\ln\frac{\sqrt[3]{d_1 d_2 d_3}}{r^1} = \frac{1.32 \times 10}{2}$$

$$\frac{\sqrt[3]{d_1 d_2 d_3}}{r^1} = e^{6.6}$$

$$\sqrt[3]{d_1 d_2 d_3} = e^{6.6} \times 4.67 \times 10^{-3}$$

$$(d_1 d_2 d_3)^{\frac{1}{3}} = 3.432$$

$$d_1 d_2 d_3 = (3.432)^3$$

But $d_1 = d$, $d_2 = d$ & $d_3 = 2d$ $\left[\because \text{Horizontal Configuration}\right]$

$$2d^3 = (3.432)^3$$

$$2d^3 = 40.455$$

$$d^3 = 20.227$$

$$d = (20.227)^{\frac{1}{3}}$$

$$d = 2.72m$$

5. Calculate the loop inductance per km of a 1 - ϕ tr. line comprising of two parallel conductors one meter apart and 1.25cm dia. Also calculate the reactance of the tr. line. Frequency is 50 Hz.

Sol: Inductance of 1 - ϕ 2 wire line $L = 4 \times 10^{-7} \ln \dfrac{d}{r^1}$ H/m

Spacing between the conductors $d = 1m = 10cm$

Diameter of the conductor $D = 1.25cm$

Radius of the conductor $r = \dfrac{D}{2} = \dfrac{1.25}{2} = 0.625cm$

GMR of the conductor $r^1 = 0.7788 \times 0.625$

$\qquad = 0.4875$ cm

Inductance of 1 - ϕ 2 wire line $= 4 \times 10^{-7} \ln \dfrac{d}{r^1}$

$\qquad = 4 \times 10^{-7} \ln \dfrac{100}{0.4875}$

$\qquad = 2.129 \ mH/Km$

Reactance of the line $X_L = 2\pi + L$

$$= 2\pi \times 50 \times 2.129 \times 10^{-3}$$

$$= 0.67\Omega$$

1.12 Inductance of 3ϕ Line with Double Circuit

It is usual practice to run 3ϕ transmission lines with more than one circuit in parallel on the same towers, because it gives greats reliability and higher transmission capacity. If such circuits are so widely separated that the mutual inductance between them becomes negligible. The inductance of a equivalent single circuit would be half of each of the individual circuits considered alone. But in actual practice the separation is not very wide and the mutual inductance is not negligible.

It is desirable to have a configuration that provides minimum inductance so as to have max. transmission capacity. This is possible only with low GMD and high GMR. Therefore the individual conductors of a phase widely separated to provide high GMR and the distance between the phases small to give low GMD. Thus in the case of a double circuit in vertical formation, the arrangement of conductors would be shown in Fig.

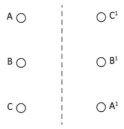

Fig. Arrangement of conductors in a double circuit 3ϕ line.

1.12.1 Inductance of a 3ϕ Double Circuit Line with Symmetrical Spacing

Consider a 3ϕ double circuit connected in parallel conductors A, B, C forming one circuit and conductors A^1, B^1, C^1 forming the other one as shown in Fig.

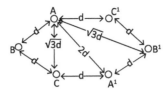

Fig. 1.19

Flux linkages of phase 'A' conductors,

$$\psi_A = 2\times 10^{-7}\left[I_A\left(ln\frac{1}{r^1}+ln\frac{1}{2d}\right)+I_B\left(ln\frac{1}{d}+ln\frac{1}{\sqrt{3}d}\right)+I_C\left(ln\frac{1}{\sqrt{3}d}+ln\frac{1}{d}\right)\right]$$

$$= 2\times 10^{-7}\left[I_A\, ln\frac{1}{2dr^1}+(I_B+I_C)ln\frac{1}{\sqrt{3}d^2}\right]$$

$$- 2\times 10^{-7}\left[I_A\, ln\frac{1}{2dr^1}-I_A\, ln\frac{1}{\sqrt{3}d^2}\right][\because I_A+I_B+I_C=0]$$

$$= 2\times 10^{-7} I_A\, ln\frac{\sqrt{3}d^2}{2dr^1}$$

$$\psi_A = 2\times 10^{-7}.\ I_A.\, ln\frac{\sqrt{3}d}{2r^1}\ \text{wb-turn/m}$$

Inductance of conductor A, $L_A = \dfrac{\psi_A}{I_A} = 2\times 10^{-7}\, ln\dfrac{\sqrt{3}d}{2r^1}$

$$L_A = 2\times 10^{-7}\, ln\frac{\sqrt{3}d}{2r^1}\ \text{H/m}$$

Inductance of remaining conductors wills also same as L_A. this is due to the fact that the conductors of different phases are symmetrically placed.

Since conductors are electrically in parallel. Inductance of each phase is, $=\dfrac{1}{2}L_A$

$$L = \frac{1}{2}\times 2\times 10^{-7}\, ln\frac{\sqrt{3}d}{2r^1}$$

$$L = 10^{-7}\, ln\frac{\sqrt{3}d}{2r^1}\ \text{H/m}$$

1.12.2 Inductance of a 3φ Double Circuit with Unsymmetrical Spacing But Transposed

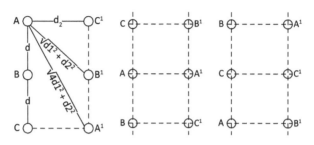

Fig. 1.20

Consider a 3φ double circuit connect in parallel – conductors A, B and C forming one circuit and conductors A^1, B^1 and forming the other one as shown in Fig.

Since the conductors are thoroughly transposed, the conductor's situations in the transposition cycle would be as shown in Fig.

Flux linkages with conductor 'A' in position (i)

$$\psi_{A_1} = 2 \times 10^{-7} \left[I_A \left(\ln \frac{1}{r^1} + \ln \frac{1}{\sqrt{4d_1^2 + d_2^2}} \right) + I_B \left(\ln \frac{1}{d_1} + \ln \frac{1}{\sqrt{d_1^2 + d_2^2}} \right) \right.$$

$$\left. + I_C \left(\ln \frac{1}{2d_1} + \ln \frac{1}{\sqrt{d_1^2 + d_2^2}} \right) \right]$$

Similarly, flux linkages with conductor A in position (ii) & (iii)

$$\psi_{A_2} = 2 \times 10^{-7} \left[I_A \left(\ln \frac{1}{r^1} + \ln \frac{1}{d_2} \right) + I_B \left(\ln \frac{1}{d_1} + \ln \sqrt{d_1^2 + d_2^2} \right) \right.$$

$$\left. + I_C \left(\ln \frac{1}{d_1} + \ln \frac{1}{\sqrt{d_1^2 + d_2^2}} \right) \right]$$

$$\psi_{A_3} = 2 \times 10^{-7} \left[I_A \left(\ln \frac{1}{r^1} + \ln \frac{1}{\sqrt{4d_1^2 + d_2^2}} \right) + I_B \left(\ln \frac{1}{2d_1} + \ln \frac{1}{d_2} \right) \right.$$

$$\left. + I_C \left(\frac{1}{d_1} + \ln \frac{1}{\sqrt{d_1^2 + d_2^2}} \right) \right]$$

Because of transposition effect,
Average flux linkages with conductor A,

$$\psi_A = \frac{\psi_{A_1} + \psi_{A_2} + \psi_{A_3}}{3}$$

$$\psi_A = \frac{2 \times 10^{-7}}{3} \left[I_A \left(3\ln\frac{1}{r^1} + \ln\frac{1}{d_2} + \ln\frac{1}{4d_1^2 + d_2^2} \right) \right.$$

$$+ I_B \left(2\ln\frac{1}{d_1} + \ln\frac{1}{2d_1} + \ln\frac{1}{d_2} + \ln\frac{1}{d_1^2 + d_2^2} \right)$$

$$\left. + I_C \left(\ln\frac{1}{2d_1} + 2\ln\frac{1}{d_1} + \ln\frac{1}{d_2} + \ln\frac{1}{d_1^2 + d_2^2} \right) \right]$$

$$\psi_A = \frac{2 \times 10^{-7}}{3} \left[I_A \left(3\ln\frac{1}{r^1} + \ln\frac{1}{d_2} + \ln\frac{1}{4d_1^2 + d_2^2} \right) \right.$$

$$\left. + (I_B + I_C) \left(\ln\frac{1}{2d_1} + \ln\frac{1}{d_2} + 2\ln\frac{1}{d_1} + \ln\frac{1}{d_1^2 + d_2^2} \right) \right]$$

$$[\because I_A + I_B + I_C = 0]$$

$$\psi_A = \frac{2 \times 10^{-7}}{3} \left[I_A \left(3\ln\frac{1}{r^1} + \ln\frac{1}{d_2} + \ln\frac{1}{4d_1^2 + d_2^2} \right) \right.$$

$$\left. - I_A \left(\ln\frac{1}{2d_1} + \ln\frac{1}{d_2} + 2\ln\frac{1}{d_1} + \ln\frac{1}{d_1^2 + d_2^2} \right) \right]$$

$$\psi_A = \frac{2 \times 10^{-7}}{3} \cdot I_A \cdot \left[\ln\frac{(d_1^2 + d_2^2) \cdot d_2 \cdot 2d_1^3}{(r^1)^3 \cdot d_2 \cdot (4d_1^2 + d_2^2)} \right]$$

$$= \frac{2 \times 10^{-7}}{3} \cdot I_A \cdot \left[\ln\frac{2d_1^3 (d_1^2 + d_2^2)}{(r^1)^3 \cdot (4d_1^2 + d_2^2)} \right]$$

$$= 2 \times 10^{-7} \cdot I_A \left[\ln\frac{(2d_1^3)^{\frac{1}{3}} \cdot (d_1^2 + d_2^2)^{\frac{1}{3}}}{(r^1)^{\frac{1}{3}} \cdot (4d_1^2 + d_2^2)^{\frac{1}{3}}} \right]$$

$$= 2\times 10^{-7}.I_A\left[\ln\frac{(2)^{\frac{1}{3}}.d_1.\left(d_1^2+d_2^2\right)^{\frac{1}{3}}}{r^1.\left(4d_1^2+d_2^2\right)^{\frac{1}{3}}}\right]$$

$$\psi_A = 2\times 10^{-7}.\ I_A.ln2^{\frac{1}{3}}.\frac{d_1}{r_1}.\frac{d_1^2+d_2^2}{\left(4d_1^2+d_2^2\right)^{\frac{1}{3}}}\ \text{wb-turn/m}$$

Inductance $L_A = \dfrac{\psi_A}{I_A}$

$$L_A = 2\times 10^{-7}.ln(2)^{\frac{1}{3}}.\frac{d_1}{r_1}.\frac{\left(d_1^2+d_2^2\right)^{\frac{1}{3}}}{\left(4d_1^2+d_2^2\right)^{\frac{1}{3}}}$$

Inductance of each phase $L = \dfrac{1}{2}L_A$

$$L = \frac{1}{2}\times 2\times 10^{-7}.ln(2)^{\frac{1}{3}}.\frac{d_1}{r_1}.\left(\frac{d_1^2+d_2^2}{4d_1^2+d_2^2}\right)^{\frac{1}{3}}$$

$$L = 2\times 10^{-7}.ln(2)^{\frac{1}{3}}.\frac{d_1}{r_1}.\left(\frac{d_1^2+d_2^2}{4d_1^2+d_2^2}\right)^{\frac{1}{6}}$$

If distance 'd_2' is too large as compased to 'd_1', $\left(\dfrac{d_1^2+d_2^2}{4d_1^2+d_2^2}\right)^{\frac{1}{6}}$

Would tend to be unity and inductance per phase,

$$L = 2\times 10^{-7}\ ln(2)^{\frac{1}{6}}.\left(\frac{d_1}{r_1}\right)^{\frac{1}{2}}$$

$$L = 0.2\ ln(2)^{\frac{1}{6}}.\left(\frac{d_1}{r_1}\right)^{\frac{1}{2}}\ \text{mH/Km}$$

1.13 Capacitance of a Transmission Line

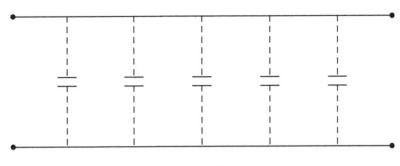

Fig. 1.21

Any two conductors separated by an insulating medium constitute a 'condenses' or a 'capacitor'. In case of an overhead line two conductors form the two plates of a capacitor and the air between the conductors behave as the dielectric medium. Thus an overhead line can be assumed to have capacitance between conductors throughout the length of the line. The capacitance is uniformly distributed over the total length of the line is shown in Fig.

When an alternating potential difference is applied across a transmission line as shown in Fig. it draws a leading current, even when supplying no load. This leading current is in quadrature with the applied voltage and is termed as charging current. The charging current is due to the capacitance effect between the conductors of the line and is not in any way dependent on the load. The strength of the charging current depends upon the voltage of transmission, the capacitance of the line and the frequency of ac supply is given as.

Charging current, $I_C = 2\pi f c V$

If the capacitance of an overhead line is high, the line draws more charging current which compensates or cancels the lagging component of the load. Hence resultant flows in the line is reduced. The reduction in resultant current flowing in the line results,

 i. Reduction of line losses and so increase of tr. efficiency.
 ii. Reduction in voltage drop or improvement of voltage regulation
 iii. Increase the load capacity & improved P.f.

1.13.1 Potential at a Charged Single Conductor

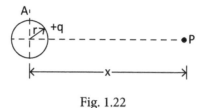

Fig. 1.22

Consider a long straight cylindrical conductor 'A' of radius 'r' meters and having a charge of 'q' coulombs per meter of its length.

The electric field intensity at a distance 'x' from the centre of conductor,

$$E = \frac{q}{2\pi \epsilon_0 \epsilon_r x} \text{ V/m}$$

Taking air as medium, i.e. $\epsilon_r = 1$

$$E = \frac{q}{2\pi \epsilon_0 x} \text{ V/m}$$

The potential difference between conductors A and infinity distant neutral plane will be equal to work done in bringing a unit +ve charge against 'E' from infinity to conductor surface and is given as,

$$V_A = \int_r^\infty \frac{q}{2\pi \epsilon_0 x} dx \Rightarrow \frac{q}{2\pi \epsilon_0} \int_r^\infty \frac{1}{x} dx$$

$$\therefore V_A = \int_r^\infty \frac{1}{x} dx$$

1.13.2 Potential at a Charged Conductor in a Group of Charged Conductors

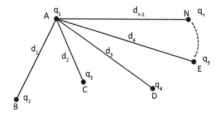

Fig. 1.23

Consider a group of long straight conductors A, B, C, D, E, N having charges of $q_1, q_2, q_3, q_4, q_5, \ldots\ldots q_x$ coulombs per meter respectively.

Potential of conductor A due to its own charge q_1,

$$= \int_r^\infty \frac{q_1}{2\pi\epsilon_0 x} \cdot \frac{1}{x} dx \text{ volts}$$

Potential of conductor A due to its own charge q_2,

$$= \int_{d_1}^\infty \frac{q_2}{2\pi\epsilon_0 x} \cdot \frac{1}{x} dx \text{ volts}$$

Since the field due to the charge q_2 extends from infinity up to a distance d_1 from conductor 'A'.

Similarly potential of conductor A, due to charge q_3 of conductor 'C' at a distance of 'd_2' meters from conductor A.

$$= \int_{d_2}^\infty \frac{q_3}{2\pi\epsilon_0 x} \cdot \frac{1}{x} dx \text{ volts}$$

Potential of conductor A due to charge q_n of conductor 'N' placed at a distance of d_{n-1} meters from conductor 'A',

$$= \int_{d_{n-1}}^\infty \frac{q_n}{2\pi\epsilon_0 x} \cdot \frac{1}{x} dx \text{ volts}$$

So, overall potential difference between conductor A and infinite distance neutral plane,

$$V_{AN} = \int_r^\infty \frac{q_1}{2\pi\epsilon_0 x} \cdot \frac{1}{x} dx + \int_{d_1}^\infty \frac{q_2}{2\pi\epsilon_0 x} \cdot \frac{1}{x} dx + \int_{d_2}^\infty \frac{q_3}{2\pi\epsilon_0 x} \cdot \frac{1}{x} dx$$

$$+ \ldots + \int_{d_{n-1}}^\infty \frac{q_n}{2\pi\epsilon_0 x} \cdot \frac{1}{x} dx$$

$$= \frac{1}{2\pi\epsilon_0} \Big[q_1(\ln\infty - \ln r) + q_2(\ln\infty - \ln d_1) + q_3(\ln\infty - \ln d_2)$$

$$+ \ldots + q_n(\ln\infty - \ln d_{n-1}) \Big]$$

$$V_{AN} = \frac{1}{2\pi\epsilon_0} \Big[q_1 \ln\frac{1}{r} + q_2 \ln\frac{1}{d_1} + q_3 \ln\frac{1}{d_2} + \ldots + \ln\frac{1}{d_{n-1}}$$

$$+ \ln\infty (q_1 + q_2 + \ldots + q_n) \Big]$$

Assuming balanced load,

i.e., $q_1 + q_2 + q_3 + \ldots + q_n = 0$

$$V_{AN} = \frac{1}{2\pi\epsilon_0}\left[q_1 \ln\frac{1}{r} + q_2 \ln\frac{1}{d_1} + q_3 \ln\frac{1}{d_2} + \ldots + \ln\frac{1}{d_{n-1}}\right] \text{volts}$$

1.13.3 Capacitance of a 1φ2 Wire Overhead Line

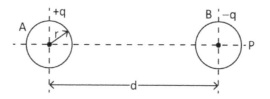

Fig. 1.24

Consider a single phase overhead line with two parallel conductors each of radius 'r' meter placed at a distance of 'd' meter in air. It is assumed that the distance 'd' b/h the condwuctors is large in comparison to the radius of the conductors. Therefore, the density of charge on either conductor will be practically unaffected by the charge on the other conductor and will be uniform throughout the length. It is assumed that the charge +1 conductor on conductor 'A' and −1 Coulombe on conductor 'B' are concentrated at the centre of the two conductor which are separated from each other by 'd' meter.

Potential difference between conductor A and neutral infinite plane,

$$V_{AN} = \int_r^\infty \frac{q}{2\pi\epsilon_0}\frac{1}{x}dx + \int_d^\infty \frac{-q}{2\pi\epsilon_0}\frac{1}{x}dx$$

$$= \frac{q}{2\pi\epsilon_0}(\ln\infty - \ln r) - \frac{q}{2\pi\epsilon_0}(\ln\infty - \ln r - \ln\infty + \ln d)$$

$$V_{AN} = \frac{q}{2\pi\epsilon_0}\ln\left(\frac{d}{r}\right)$$

Similarly P.D between conductor 'B' and infinite neutral plane,

$$V_{BN} = \int_r^\infty \frac{-q}{2\pi\epsilon_0}\frac{1}{x}dx + \int_d^\infty \frac{q}{2\pi\epsilon_0}\frac{1}{x}dx$$

$$V_{BN} = \frac{-q}{2\pi\epsilon_0} \ln\left(\frac{d}{r}\right)$$

P.D. between conductor A and B

$$V_{AB} = V_{AN} - V_{BN}$$

$$= \frac{q}{2\pi\epsilon_0}\ln\left(\frac{d}{r}\right) - \left(\frac{-q}{2\pi\epsilon_0}\right)\ln\left(\frac{d}{r}\right)$$

$$V_{AB} = \frac{q}{2\pi\epsilon_0}\ln\left(\frac{d}{r}\right)$$

P.D. between conductor A and B,

$$V_{AB} = V_{AN} - V_{BN}$$

$$= \frac{q}{2\pi\epsilon_0}\ln\left(\frac{d}{r}\right) - \left(\frac{-q}{2\pi\epsilon_0}\right)\ln\left(\frac{d}{r}\right)$$

$$V_{AB} = \frac{q}{\pi\epsilon_0}\ln\left(\frac{d}{r}\right)$$

Capacitance of the line, $C = \dfrac{q}{V_{AB}}$

$$= \frac{q}{\dfrac{q}{\pi\epsilon_0}\ln\left(\dfrac{d}{r}\right)}$$

$$c = \frac{\pi\epsilon_0}{\ln\left(\dfrac{d}{r}\right)} \text{ F/m}$$

1.14 Effect of Earth on the Capacitance of 1ϕ OHTL

The presence of earth alters the electric field of the line and also affects its capacitance. Earth may be assumed to be a perfect conductor in the form of horizontal plane of infinite extent. Therefore the electric field of charged conductor is forced to conform a presence of this equipotential surface. Such problems handled by "Method of images".

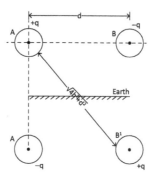

Fig. 1.25

The earth is considered to be at zero potential an it is possible only if there is an image conductor 'A' having a charge of '–1' coulombs per meter length at a depth of 'h' meters below the earth. Similarly there is a conductor 'B' below the earth having a charge of '+q' coulombs per meter length.

Potential of conductor A,

$$V_{AN} = \int_r^\infty \frac{q}{2\pi\epsilon_0} \frac{1}{x} dx + \int_d^\infty \frac{-q}{2\pi\epsilon_0} \frac{1}{x} dx + \int_{2h}^\infty \frac{-q}{2\pi\epsilon_0} \frac{1}{x} dx + \int_{\sqrt{14h^2+d^2}}^\infty \frac{q}{2\pi\epsilon_0} \frac{1}{x} dx$$

$$= \frac{q}{2\pi\epsilon_0}\left[\ln\infty - \ln r - \ln\infty + \ln d - \ln\infty + \ln 2h + \ln\infty - \ln\sqrt{4h^2+d^2}\right]$$

$$= \frac{q}{2\pi\epsilon_0}\left[\ln\frac{1}{r} + \ln d + \ln 2h - \ln\sqrt{4h^2+d^2}\right]$$

$$= \frac{q}{2\pi\epsilon_0}\left[\ln\frac{d}{r} + \ln\frac{2h}{\sqrt{4h^2+d^2}}\right]$$

$$V_{AN} = \frac{q}{2\pi\epsilon_0}\ln\frac{2hd}{r\sqrt{4h^2+d^2}}$$

Similarly potential of conductor B,

$$V_{BN} = \int_r^\infty \frac{-q}{2\pi\epsilon_0} \frac{1}{x} dx + \int_d^\infty \frac{q}{2\pi\epsilon_0} \frac{1}{x} dx + \int_{2h}^\infty \frac{q}{2\pi\epsilon_0} \frac{1}{x} dx + \int_{\sqrt{14h^2+d^2}}^\infty \frac{-q}{2\pi\epsilon_0} \frac{1}{x} dx$$

$$V_{BN} = \frac{-q}{2\pi\epsilon_0}\ln\frac{2hd}{r\sqrt{4h^2+d^2}} \text{ volts}$$

P.D. between conductor A and B,

$$V_{AB} = V_{AN} - V_{BN}$$

$$= \left[\frac{q}{2\pi\epsilon_0} \ln \frac{2hd}{r\sqrt{4h^2+d^2}} - \frac{-q}{2\pi\epsilon_0} \ln \frac{2hd}{r\sqrt{4h^2+d^2}} \right]$$

$$= \frac{q}{\pi\epsilon_0} \ln \frac{2hd}{r\sqrt{4h^2+d^2}}$$

$$V_{AB} = \frac{q}{\pi\epsilon_0} \ln \frac{d}{r\sqrt{4h^2+d^2}} \text{ volts}$$

Capacitance between conductor A and B,

$$C_{AB} = \frac{q}{V_{AB}}$$

$$= \frac{q}{\frac{q}{\pi\epsilon_0} \ln \frac{d}{r\sqrt{4h^2+d^2}}}$$

$$C_{AB} = \frac{\pi\epsilon_0}{\ln \frac{d}{r\sqrt{1+\frac{d^2}{4h^2}}}} \text{ F/m}$$

1.15 Capacitance of 3φ Single Circuit Equilateral Spacing

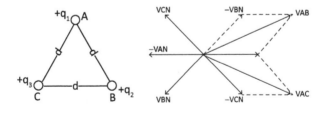

Fig. 1.26

Consider 3φ line with conductors A, B and C each of radius 'r' meter and conductors are spaced equilaterally with a diameter of 'd' meter.

We will assume that there are no other charged surfaces in vicinity (Neglecting the ground effect), so that the sum of charges q_1, q_2 and q_3 is zero.

The voltage drop between any two conductors is sum of the voltage drops due to each charged conductors. So we can write phase voltage from A to B and A to C as

$$V_{AB} = V_{AN} - V_{BN}$$

$$V_{AN} = \frac{1}{2\pi\epsilon_0}\left[q_1 \ln\frac{1}{r} + q_2 \ln\frac{1}{d} + q_3 \ln\frac{1}{d}\right]$$

$$V_{BN} = \frac{1}{2\pi\epsilon_0}\left[q_1 \ln\frac{1}{d} + q_2 \ln\frac{1}{r} + q_3 \ln\frac{1}{d}\right]$$

$$V_{AB} = \frac{1}{2\pi\epsilon_0}\left[q_1 \ln\frac{d}{r} + q_2 \ln\frac{r}{d}\right]$$

$$V_{AC} = V_{AN} - V_{CN}$$

$$V_{AN} = \frac{1}{2\pi\epsilon_0}\left[q_1 \ln\frac{1}{r} + q_2 \ln\frac{1}{d} + q_3 \ln\frac{1}{d}\right]$$

$$V_{CN} = \frac{1}{2\pi\epsilon_0}\left[q_1 \ln\frac{1}{d} + q_2 \ln\frac{1}{d} + q_3 \ln\frac{1}{r}\right]$$

$$V_{AC} = \frac{1}{2\pi\epsilon_0}\left[q_1 \ln\frac{d}{r} + q_3 \ln\frac{r}{d}\right]$$

$$V_{AB} + V_{AC} = \frac{1}{2\pi\epsilon_0}\left[q_1 \ln\frac{d}{r} + q_2 \ln\frac{r}{d} + q_1 \ln\frac{d}{r} + q_3 \ln\frac{r}{d}\right]$$

$$= \frac{1}{2\pi\epsilon_0}\left[2q_1 \ln\frac{d}{r} + (q_2 + q_3)\ln\frac{r}{d}\right]$$

$$= \frac{1}{2\pi\epsilon_0}\left[2q_1 \ln\frac{d}{r} - q_1 \ln\frac{r}{d}\right]$$

$$= \frac{1}{2\pi\epsilon_0}\left[2q_1 \ln\frac{d}{r} + q_1 \ln\frac{d}{r}\right]$$

$$V_{AB} + V_{AC} = \frac{3q_1}{2\pi\epsilon_0}\ln\frac{d}{r}$$

From the phasor, it is clear that $V_{AB} + V_{AC} = 3 V_{AN}$

$$\therefore 3 V_{AN} = \frac{3q_1}{2\pi\epsilon_0} \ln\frac{d}{r}$$

$$V_{AN} = \frac{q_1}{2\pi\epsilon_0} \ln\frac{d}{r}$$

Capacitance of conductor A, $C_A = \dfrac{q_1}{V_{AN}}$

$$C_A = \frac{q_1}{\dfrac{q_1}{2\pi\epsilon_0}\ln\dfrac{d}{r}} \Rightarrow C_A = \frac{2\pi\epsilon_0}{\ln\dfrac{d}{r}} \text{ F/m}$$

Since conductors are equilaterally spaced, $C_A = C_B = C_C = C$

$$\therefore C = \frac{2\pi\epsilon_0}{\ln\dfrac{d}{r}} \text{ F/m}$$

1.16 Capacitance of 3 – φ Double Circuit with Symmetrical Configuration

Consider a 3-phase double circuit connected in parallel – conductors A, B and C forming one circuit and conductors A^1, B^1 and C^1 forming another circuit (conductors symmetrically spaced).

Let the charge over conductors A, B and C be q_1, q_2 and q_3 coulombs per metre length. Then charge over conductors A^1, B^1 & C^1 will obviously be q_1, q_2 and q_3 coulombs per metre length and $q_1 + q_2 + q_3 = 0$.

Potential of conductors A w.r.t neutral infinite plane

$$V_{AN} = \frac{1}{2\pi\epsilon_0}\left[q_1\int_r^\infty \frac{dx}{x} + q_2\int_d^\infty \frac{dx}{x} + q_3\int_{\sqrt{3}d}^\infty \frac{dx}{x} + q_1\int_{2d}^\infty \frac{dx}{x} + q_2\int_{\sqrt{3}d}^\infty \frac{dx}{x} + q_3\int_d^\infty \frac{dx}{x}\right]$$

$$= \frac{1}{2\epsilon}\left[q_1\log_e\infty - q_1\log_e r + q_2\log_e\infty - q_2\log_e d - q_3\log_e\sqrt{3}d + q_3\log_e\infty\right.$$
$$\left. + q_1\log_e\infty - q_1\log_e 2d + q_2\log_e\infty - q_2\log_e\sqrt{3}d + q_3\log_e\infty - q_3\log_e\frac{1}{\sqrt{}}\right]$$

$$= \frac{1}{2\pi\epsilon_0}\left[2(q_1+q_2+q_3)\log_e \infty + q_1 \log_e \frac{1}{2dr} + q_2 \log_e \frac{1}{\sqrt{3d^2}} + q_3 \log_e \frac{1}{\sqrt{3d^2}}\right]$$

$$= \frac{1}{2\pi\epsilon_0}\left[q_1 \log_e \frac{1}{2dr} + q_2 \log_e \frac{1}{\sqrt{3d^2}} + q_3 \log_e \frac{1}{\sqrt{3d^2}}\right]$$

$$[\because q_1+q_2+q_3 = 0]$$

$$= \frac{1}{2\pi\epsilon_0}\left[q_1 \log_e \frac{1}{2dr} + (q_2+q_3)\log_e \frac{1}{\sqrt{3d^2}}\right]$$

$$= \frac{1}{2\pi\epsilon_0}\left[q_1 \log_e \frac{1}{2dr} - q_1 \log_e \frac{1}{\sqrt{3d^2}}\right] \qquad [\because q_2+q_3 = -q_1]$$

$$= \frac{q_1}{2\pi\epsilon_0}\left[\log_e \frac{\sqrt{3d^2}}{2dr}\right]$$

$$= \frac{q_1}{2\pi\epsilon_0}\left[\log_e \frac{\sqrt{3}d}{2r}\right] \text{ volts}$$

Capacitance of conductor A to neutral

$$C_{AN} = \frac{q_1}{V_{AN}} = \frac{q_1}{\frac{q_1}{2\pi\epsilon_0}\log_e \frac{\sqrt{3}d}{2r}} = \frac{2\pi\epsilon_0}{\log_e \frac{\sqrt{3}d}{2r}} \text{ F/m}$$

Similarly expressions for C_{BN} and C_{CN} can be obtained and we have

$$C_{AN} = C_{BN} = C_{CN} = C_N = \frac{2\pi\epsilon_0}{\log_e \frac{\sqrt{3}d}{2r}} \text{ F/m} \rightarrow (1)$$

This is because the conductors of different phases are symmetrically placed.

Equation (1) gives the capacitance of conductor A alone, whereas there are two conductors per phase A and A^1. Therefore the capacitance of the system per phase will be twice of the capacitance of one conductor to neutral i.e.,

Capacitance per phase, $C = 2 C_N = \dfrac{4\pi\epsilon_0}{\log_e \dfrac{\sqrt{3}d}{2r}}$ F/m

1.17 Capacitance of 3 − φ Single Circuit Overhead Line (Unsymmetrical Configuration)

For an transposed unsymmetrical 3 − φ line the capacitances between conductor to neutral of the 3 conductors are different. Suppose that the line is, as shown in Fig, and that voltages V_A, V_B, V_C are applied to the conductors with the result that the charges per meter length are q_1, q_2, q_3 respectively.

$$V_{AN} = \frac{q_1}{2\pi\epsilon_0}\int_r^\infty \frac{dx}{x} + \frac{q_2}{2\pi\epsilon_0}\int_{d_1}^\infty \frac{dx}{x} + \frac{q_3}{2\pi\epsilon_0}\int_{d_3}^\infty \frac{dx}{x}$$

$$= \frac{1}{2\pi\epsilon_0}\left[q_1\log_e\infty - q_1\log_e r + q_2\log_e\infty - q_2\log_e d_1 + q_3\log_e\infty - q_3\log_e d_3\right]$$

$$= \frac{1}{2\pi\epsilon_0}\left[q_1\log_e\frac{1}{r} + q_2\log_e\frac{1}{d_1} + q_3\log_e\frac{1}{d_3} + (q_1+q_2+q_3)\log_e\infty\right]$$

$$V_{AN} = \frac{1}{2\pi\epsilon_0}\left[q_1\log_e\frac{1}{r} + q_2\log_e\frac{1}{d_1} + q_3\log_e\frac{1}{d_3}\right] \text{ volts} \quad \rightarrow (1)$$

$$[\because q_1+q_2+q_3=0]$$

Similarly, $V_{BN} = \dfrac{1}{2\pi\epsilon_0}\left[q_2\log_e\dfrac{1}{r} + q_1\log_e\dfrac{1}{d_1} + q_3\log_e\dfrac{1}{d_2}\right]$ volts $\quad \rightarrow (2)$

& $V_{CN} = \dfrac{1}{2\pi\epsilon_0}\left[q_3\log_e\dfrac{1}{r} + q_1\log_e\dfrac{1}{d_3} + q_2\log_e\dfrac{1}{d_2}\right]$ volts $\quad \rightarrow (3)$

Substituting $q_3 = -(q_1+q_2)$ in Equations (2) & (3) we have

$$V_{AN} = \frac{1}{2\pi\epsilon_0}\left[q_1\log_e\frac{1}{r} + q_2\log_e\frac{1}{d_1} - (q_1+q_2)\log_e\frac{1}{d_3}\right]$$

$$= \frac{1}{2\pi\epsilon_0}\left[q_1\log_e\frac{d_3}{r} + q_2\log_e\frac{d_3}{d_1}\right] \text{ volts} \quad \rightarrow (4)$$

and $V_{BN} = \dfrac{1}{2\pi\epsilon_0}\left[q_2\log_e\dfrac{1}{r} + q_1\log_e\dfrac{1}{d_1} - (q_1+q_2)\log_e\dfrac{1}{d_2}\right]$ volts

$$= \frac{1}{2\pi\epsilon_0}\left[q_1\log_e\frac{d_2}{d_1} + q_2\log_e\frac{d_2}{r}\right] \text{ volts} \quad \rightarrow (5)$$

Multiplying Eq. (4) by $\log_e \dfrac{d_2}{r}$ and Eq. (5) by $\log_e \dfrac{d_3}{d_1}$ we have

$$V_{AN} \log_e \frac{d_2}{r} = \frac{1}{2\pi\epsilon_0}\left[q_1 \log_e \frac{d_3}{r} \cdot \log_e \frac{d_2}{r} + q_2 \log_e \frac{d_3}{d_1} \log_e \frac{d_2}{r} \right] \quad \rightarrow (6)$$

and $V_{BN} \dfrac{d_3}{r} = \dfrac{1}{2\pi\epsilon_0}\left[q_1 \log_e \dfrac{d_2}{d_1} \cdot \log_e \dfrac{d_3}{d_1} + q_2 \log_e \dfrac{d_2}{r} \cdot \log_e \dfrac{d_3}{d_1} \right] \quad \rightarrow (7)$

subtracting Eq. (7) from Eq. (6) we have,

$$V_{AN} \log_e \frac{d_2}{r} - V_{BN} \log_e \frac{d_3}{d_1} = \frac{q_1}{2\pi\epsilon_0}\left[\log_e \frac{d_3}{r} \log_e \frac{d_2}{r} - \log_e \frac{d_2}{d_1} \log_e \frac{d_3}{d_1} \right]$$

(or)

$$q_1 = 2\pi\epsilon_0 \frac{V_{AN} \log_e \dfrac{d_2}{r} - V_{BN} \log_e \dfrac{d_3}{d_1}}{\log_e \dfrac{d_3}{r} \log_e \dfrac{d_2}{r} - \log_e \dfrac{d_2}{d_1} \log_e \dfrac{d_3}{d_1}} \quad \rightarrow (8)$$

Capacitance of conductor A to neutral,

$$C_{AN} = \frac{q_1}{V_{AN}} = 2\pi\epsilon_0 \frac{\log_e \dfrac{d_3}{r} - \dfrac{V_{BN}}{V_{AN}} \log_e \dfrac{d_3}{d_1}}{\log_e \dfrac{d_3}{r} \log_e \dfrac{d_2}{r} - \log_e \dfrac{d_2}{d_1} \cdot \log_e \dfrac{d_3}{d_1}} \text{ f/m} \quad \rightarrow (9)$$

Similarly capacitance of conductor B to neutral,

$$C_{BN} = \frac{q_2}{V_{BN}} = \frac{2\pi\epsilon_0 \log_e \dfrac{d_3}{r} - \dfrac{V_{CN}}{V_{BN}} \log_e \dfrac{d_1}{d_2}}{\log_e \dfrac{d_1}{r} \log_e \dfrac{d_3}{r} - \log_e \dfrac{d_3}{d_2} \cdot \log_e \dfrac{d_1}{d_2}} \text{ f/m}$$

$$C_{CN} = 2\pi\epsilon_0 \frac{\log_e \dfrac{d_1}{r} - \dfrac{V_{AN}}{V_{CN}} \log_e \dfrac{d_2}{d_3}}{\log_e \dfrac{d_2}{r} \log_e \dfrac{d_1}{r} - \log_e \dfrac{d_1}{d_3} \cdot \log_e \dfrac{d_2}{d_3}} \text{ f/m}$$

1.18 Capacitance of 3 – φ Double Circuit with Unsymmetrical Configuration

Consider conductors are arranged as shown below, corresponding to different positions in the transposition positions.

Potential of conductor A w.r.t infinite neutral plane.

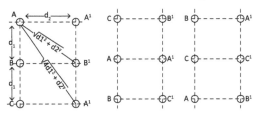

Fig. 1.27

$$V_{AN_1} = \frac{1}{2\pi\epsilon_0}\left[q_1\int_r^\infty \frac{dx}{x} + q_2\int_{d_1}^\infty \frac{dx}{x} + q_3\int_{2d_1}^\infty \frac{dx}{x} + q_1\int_{\sqrt{4d_1^2+d_2^2}}^\infty \frac{dx}{x} + q_2\int_{\sqrt{d_1^2+d_2^2}}^\infty \frac{dx}{x} \right.$$
$$\left. + q_3\int_{d_2}^\infty \frac{dx}{x}\right] \text{volts}$$

$$= \frac{1}{2\pi\epsilon_0}\left[q_1\log_e \frac{1}{r\sqrt{4d_1^2+d_2^2}} + q_2\log_e \frac{1}{d_1\sqrt{d_1^2+d_2^2}} + q_3\log_e \frac{1}{2d_1d_2}\right] \text{volts}$$

Similarly

$$V_{AN_2} = \frac{1}{2\pi\epsilon_0}\left[q_1\log_e \frac{1}{rd_2} + q_2\log_e \frac{1}{d_1\sqrt{d_1^2+d_2^2}} + q_3\log_e \frac{1}{d_1\sqrt{d_1^2+d_2^2}}\right] \text{volts and}$$

$$V_{AN_3} = \frac{1}{2\pi\epsilon_0}\left[q_1\log_e \frac{1}{r\sqrt{4d_1^2+d_2^2}} + q_2\log_e \frac{1}{2d_1d_2} + q_3\log_e \frac{1}{d_1\sqrt{d_1^2+d_2^2}}\right]$$

$$V_{AN} = \frac{V_{AN_1} + V_{AN_2} + V_{AN_3}}{3}$$

$$= \frac{1}{6\pi\epsilon_0}\left[q_1\log_e \frac{1}{r^3 d_2\left(4d_1^2+d_2^2\right)} + q_2\log_e \frac{1}{2d_1^3 d_2\left(d_1^2+d_2^2\right)}\right.$$
$$\left. + q_3\log_e \frac{1}{2d_1^3 d_2\left(d_1^2+d_2^2\right)}\right]$$

$$= \frac{1}{6\pi\epsilon_0}\left[q_1 \log_e \frac{1}{r^3 d_2 \left(4d_1^2 + d_2^2\right)} + (q_2 + q_3)\log_e \frac{1}{2d_1^3 d_2 \left(d_1^2 + d_2^2\right)}\right]$$

$$= \frac{1}{6\pi\epsilon_0}\left[q_1 \log_e \frac{1}{r^3 d_2 \left(4d_1^2 + d_2^2\right)} - q_1 \log_e \frac{1}{2d_1^3 d_2 \left(d_1^2 + d_2^2\right)}\right] \left[\because q_2 + q_3 = -q_1\right]$$

$$= \frac{q_1}{6\pi\epsilon_0}\left[q_1 \log_e \frac{2d_1^3 d_2 \left(d_1^2 + d_2^2\right)}{r^3 d_2 \left(4d_1^2 + d_2^2\right)}\right]$$

$$= \frac{q_1}{2\pi\epsilon_0} \log_e \frac{2^{\frac{1}{3}} d_1}{r}\left[\frac{d_1^2 + d_2^2}{4d_1^2 + d_2^2}\right]^{\frac{1}{3}} \text{ Volts}$$

Capacitance of conductor A

$$C_{PN} = \frac{q_1}{V_{AN}} = \frac{q}{\frac{q}{2\pi\epsilon_0} \log_e 2^{\frac{1}{3}} \frac{d_1}{r}\left[\frac{d_1^2 + d_2^2}{4d_1^2 + d_2^2}\right]^{\frac{1}{3}}}$$

$$= \frac{2\pi\epsilon_0}{\log_e 2^{\frac{1}{3}} \frac{d_1}{r}\left[\frac{d_1^2 + d_2^2}{4d_1^2 + d_2^2}\right]^{\frac{1}{3}}}$$

Similarly expressions for capacitance C_{BN} and C_{CN} can be obtained which are same as above

Capacitance per phase will be double of C_{AN} i.e., Capacitance per phase

$$C = \frac{4\pi\epsilon_0}{\log_e 2^{\frac{1}{3}} \frac{d_1}{r}\left[\frac{d_1^2 + d_2^2}{4d_1^2 + d_2^2}\right]^{\frac{1}{3}}}$$

1. **Find out capacitance of single phase line 30 km long consisting of two parallel wires of each 15mm diameter 15m. apart.**

   Given data,

   Diameter of conductor, $D = 15mm$

   Radius of conductor, $r = 7.5mm$

   Spacing between conductors, $d = 1.5m = 1500mm$

Capacitance of line $C = \dfrac{\pi \epsilon_0}{\ln\left(\dfrac{d}{r}\right)}$ f/m

$$C = \dfrac{\pi \times 8.854 \times 10^{-12}}{\ln\left(\dfrac{1500}{7.5}\right)} = 5.24 \times 10^{-12} \text{ f/m}$$

Capacitance of line for 30km,
$$C = 5.24 \times 10^{-12} \times 30 \times 10^3 = 1.57 \times 10^{-7}$$
$$\therefore C = 0.157 \, \mu f$$

2. Find out capacitance of 1 – ϕ two wire line running at a height of 'h' m. above earth. Calculate capacitance to neutral in case of 1 – ϕ line whose conductors with radius of 0.25cm are separated by 1.5m. & which are lying 7m. above ground. The line length is 50km.

Given data,
Height above the earth is 'h' m
Radius of conductor $r = 0.25cm$
Spacing b/w conductors $d = 1.5m = 150cm$
Length of the line is 50km
Height of conductor above earth, $h = 7m = 700m$
Capacitance,

$$C = \dfrac{\pi \epsilon_0}{\ln\left(\dfrac{d}{r\sqrt{1 + \dfrac{(150)^2}{4(700)^2}}}\right)}$$

$$C = \dfrac{\pi \times 8.854 \times 10^{-12}}{\ln\left(\dfrac{150}{0.25\sqrt{1 + \dfrac{(150)^2}{4(700)^2}}}\right)}$$

$C = 4.35 \times 10^{-12}$ f/m

Capacitance of line for 50km,

$C = 4.35 \times 10^{-12} \times 50 \times 10^3 = 2.175 \times 10^{-7}$ F

$\therefore C = 0.217 \, \mu f$

3. **Calculate capacitance of 100km long 3ϕ 50Hz over head transmission line of 3 conductors each of diameter 2cm of spacing 2.5m. at the comers of an equilateral triangle.**

Given data,

Length of the line $= 100 km$

Diameter of conductor $D = 2cm$

Radius of conductor $= \dfrac{2}{2} = 1cm$

Spacing b/w the conductors, $d = 2.5cm = 250cm$

Capacitance,

$$C = \dfrac{2\pi\epsilon_0}{\ln\left(\dfrac{d}{r}\right)} = \dfrac{2\pi \times 8.854 \times 10^{-12}}{\ln\left(\dfrac{2.5 \times 100}{1}\right)}$$

$C = 1.007 \times 10^{-11}$ F

$C = 10.07 \times 10^{-12}$ F/m

Capacitance of line for 100km.

$C = 10.07 \times 10^{-12} \times 100 \times 10^3$

$C = 1.007 \, \mu f$

4. **The 2cm diameter conductor of a 3ϕ 3-wire transmission line are situated at corners of a triangle of sides 3.5m, 5m, 8m. Find capacitance of line is transposed.**

Given data,

Spacing b/w conductors,

$d_1 = 3.5m = 350cm$

$d_2 = 5m = 500cm$

$d_3 = 8m = 800cm$

Diameter of conductor, $d = 2m$

Radius of conductor, $r = \dfrac{d}{2} = 1cm = 0.01m$

Capacitance of conductor,

$$C = \dfrac{2\pi\epsilon_0}{\ln\dfrac{\sqrt[3]{d_1 d_2 d_3}}{r}} \; f/m$$

$$= \dfrac{2\pi\epsilon_0}{\ln\dfrac{\sqrt[3]{3.5\times 5\times 8}}{0.01}}$$

$C = 8.89 \times 10^{-12} \; f/m$

Capacitance for 1km,

$C = 8.89 \times 10^{-12} \times 1000 = 0.008 \; \mu f$

5. **A 3φ, 132KV, 50Hz, OHTL has steel core aluminium conductor of equivalent copper area of 1.5 cm² & effective diameter of 39.2mm spaced equilaterally 8m. part. Calculate line constants per km. length of line. Resistivity of copper is 1.73μΩ-cm.**

Given data,

Diameter $= 39.2mm$

Area of conductor $= 1.5 cm^2$

Radius of conductor $= 19.6mm = 0.0196m$

Spacing b/w conductors, $d = 8m$

Resistivity, $\rho = 1.73 \; \mu\Omega$-cm

Frequency $f = 50Hz$

(i) Inductance of line, $L = 0.2 \ln\left(\dfrac{d}{r^1}\right)$ mH/Km

$$L = 0.2 \ln\left(\dfrac{8}{0.7788 \times 0.0196}\right) = 1.25 mH/Km$$

$L = 1.25 \; mH/Km$

(ii) Capacitance of line, $C = \dfrac{2\pi\epsilon_0}{\ln\left(\dfrac{d}{r}\right)} \; f/m$

$$C = \dfrac{2\pi \times 8.854 \times 10^{-12}}{\ln\left(\dfrac{8}{0.0196}\right)} = 9.25 \times 10^{-12} \; f/m$$

$\therefore C = 0.00925 \ \mu f / Km$

(iii) Resistance of line, $R = \dfrac{\rho L}{a}$

$R = \dfrac{1.73 \times 10^{-6} \times 1000 \times 100}{1.5}$

$R = 0.1153 \ \Omega / Km$

6. A 3ϕ 50Hz 132KV OHTL has conductors placed in a horizontal plane. 4.56 m. apart conductor dia is 22.4 mm. If line current / phase. Assuming complete transposition.

```
        A        B        C
        O        O        O
        |← 4.56m ─*─ 4.56m →|
        |←───── 9.12m ─────→|
```

Given data,
Frequency $= 50Hz$
Line voltage $= 132 \times 10^3 V = 132KV$

Phase voltage $= \dfrac{132}{\sqrt{3}} = 76.21KV$

Spacing b/w conductors
$d_1 = d_2 = 4.56m$
$d_3 = 9.12m$

Diameter of conductor, $D = 22.4mm$
Radius of conductor, $r = 11.2mm$
$r = 0.0112m$

Length of line $= 100Km$

Capacitance $C = \dfrac{2 \pi \epsilon_0}{\ln\left(\dfrac{\sqrt[3]{d_1 d_2 d_3}}{r}\right)}$ f/m

$C = \dfrac{2 \pi \times 8.854 \times 10^{-12}}{\ln\left(\dfrac{\sqrt[3]{4.56 \times 4.56 \times 9.12}}{0.0112}\right)} = 8.91 \times 10^{-12}$ f/m

Capacitance for 100Km

$C = 8.91 \times 10^{-12} \times 100 \times 10^3$

$C = 0.89 \; \mu f$

Charging current, $I_C = 2\pi f C V$

$I_C = 2\pi \times 50 \times 0.89 \times 10^{-6} \times 76.21 \times 10^3$

$I_C = 21.308 \; A$

7. **A single 3φ line operated at 50Hz is arranged as shown in Fig. the conductor dia is 8mm & line is regularly transposed. Find inductance & capacitance per km length of line.**

Given data,

Diameter of conductor = 8mm

Radius of conductor = 4mm

$r = 4 \times 10^{-3} \; m$

Frequency $f = 50Hz$

Spacing b/w conductors,

$d_1 = 1.5m$

$d_2 = 3m$

$d_3 = 2.6m$

Inductance/km = $0.2 \ln \dfrac{\sqrt[3]{d_1 d_2 d_3}}{r^1}$

$L = 0.2 \ln \dfrac{\sqrt[3]{1.5 \times 3 \times 2.6}}{0.7788 \times 4 \times 10^{-3}}$

$L = 1.318 \; mH/Km$

Capacitance = $\dfrac{2\pi \epsilon_0}{\ln\left(\dfrac{\sqrt[3]{d_1 d_2 d_3}}{r}\right)}$

$C = \dfrac{2\pi \times 8.854 \times 10^{-12}}{\ln\left(\dfrac{\sqrt[3]{3 \times 2.6 \times 1.5}}{4 \times 10^{-3}}\right)}$

$C = 8.77 \times 10^{-12} \; f/m$

$C = 0.0087 \; \mu f$

8. **Calculate capacitance of 1 - ϕ transmission line 35km long consisting of two parallel wires each 5mm dia & 1.8m apart. Height of conductor above ground is 7.5m**

Given data,

Length of line $= 3.5 Km$

Diameter of conductor $= 5mm$

Radius $= \dfrac{5}{2} = 2.5mm = 2.5 \times 10^{-3} m$

Spacing between conductor, $d = 1.8m$

Height of conductor above ground, $h = 7.5m$

Capacitance,

$$C = \dfrac{\pi \epsilon_0}{\ln\left(\dfrac{d}{r\sqrt{1+\dfrac{d^2}{4h^2}}}\right)} = \dfrac{\pi \times 8.854 \times 10^{12}}{\ln\left(\dfrac{1.8}{2.5 \times 10^{-3}\sqrt{1+\dfrac{(1.8)^2}{4(7.5)^2}}}\right)}$$

$C = 4.23 \times 10^{-12} \, f/m$

For 35Km,

$C = 1.48 \times 10^{-7} \, f$

$C = 0.148 \, \mu f$

9. **A 1 - ϕ line constructed 13.5m. above ground & spacing between conductors being 3.9m, Radius of conductor is 1.78cm. Find capacitance of line/m. length, consider in effect of earth & neglecting it.**

Given,

Height of line above ground $h = 13.5m$

Spacing b/w conductors $d = 3.9m$

Radius of conductor $r = 1.78cm = 0.0178m$

(i) Capacitance of line $= \dfrac{\pi \epsilon_0}{\ln\left(\dfrac{d}{r}\right)}$ f/m

$C = \dfrac{\pi \times 8.854 \times 10^{-12}}{\ln\left(\dfrac{3.9}{0.0178}\right)} = 5.161 \times 10^{-12}$

Capacitance/meter length $= 5.16 \times 10^{-12}$ f/m

(ii) Capacitance of line, considering effect of earth

$C = \dfrac{\pi \epsilon_0}{\ln\left(\dfrac{d}{r\sqrt{1+\dfrac{d^2}{4h^2}}}\right)} = \dfrac{\pi \times 8.854 \times 10^{-12}}{\ln\left(\dfrac{3.9}{0.0178\sqrt{1+\dfrac{(3.9)^2}{4(13.5)^2}}}\right)}$

$C = 5.179 \times 10^{-12}$ f/m

10. Find capacitance & inductance per phase of 3φ double circuit line, conductors of which are arranged in hexagonally spaced, distance b/w conductors being 2.5m. The diameter of each conductors is 3cm. Total length of line is 120Km.

Given,
Spacing b/w conductors
$d = 2.5$m

Diameter of each conductor = 3cm
$D = 0.03$m

Radius of conductor = 0.015m

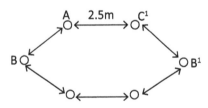

Capacitance/phase $= \dfrac{4\pi \epsilon_0}{\ln\dfrac{\sqrt{3}d}{2r}}$ f/m

$C = \dfrac{4\pi \times 8.854 \times 10^{-12}}{\ln\dfrac{\sqrt{3} \times 2.5}{2 \times 0.015}} = 2.237 \times 10^{-11}$ f/m

$C = 22.37 \times 10^{-12}$ f/m

Capacitance/Km $= 22.37 \times 10^{-12} \times 120 \times 10^3 = 2.68\,\mu f/km$

Inductance/phase $= 10^{-7} \ln\left(\dfrac{\sqrt{3}d}{2r^1}\right)$ H/m

$L = 10^{-7} \ln\left(\dfrac{\sqrt{3} \times 2.5}{2(0.7788 \times 0.015)}\right)$

$L/ph = 0.52 \times 10^{-6}$ H/m

CHAPTER 2

Performance of Short and Medium Length Transmission Lines

2.1 Introduction

The transmissionline is the main corridor in a power system. The performance of a power system depends mainly on the performance of transmission lines in the system. The important considerations in the operation of the transmission lines are voltage drop and power loss occurring in the line and efficiency of transmission.

The performance of a transmission line is governed by its four parameters – series resistance 'R' and inductance 'L', shunt Capacitance 'C' and conductance 'G'. The resistance 'R' is due to the fact that every conductor offers opposition to the flow of current. The inductance 'L' is due to the fact that the current carrying conductor is surrounded by the magnetic lines of force the capacitance of the line is due to the fact that the conductor carrying current forms a capacitor with the earth which is always at lower potential than the conductor. air between them forms a dielectric medium. The shunt conductance is mainly due to flow of leakage currents over the surface of the insulator especially during bad weather conditions.

The effect of line resistance is to cause voltage drop (IR) and power loss (I^2R) in the line. The effect of the line inductance is to cause voltage drop in the line in quadrature with the current flowing in the conductor which is equal to $2\pi fL$. The effect of line capacitance is to produce a current and it is called "Charging current which is in quadrature with voltage.

It is to be noted that both reactive drop (IX_L) and the charging current ($2\pi fCV_c$) of the line are proportional to the supply frequency and thus have for greater influence on the performance of a 50Hz line than on 25 Hz line.

In 3ϕ circuit problems, it is sufficient to compute results for one phase and subsequently predict results for the remaining two phases by exploiting 3ϕ symmetry. Although the transmission line are not spaced equilaterally and not transposed, the resulting asymmetry is slight and the phases are considered to be balanced. A 3ϕ transmission line carrying an equal load on each phase can be represented by its single phase equivalent in which inductance is computed for one phase of a balanced 3ϕ line and capacitance is computed for line to neutral.

2.2 Classification of Transmission Lines

The over head transmission lines are classified depending up on the manner in which capacitance is taken into account and length of the line.

Transmission lines having length lesser than 80km and operating voltage lower than 20 kV will fall into the category called "shorttransmission lines". Due to smaller distance and lower line voltage, the capacitance effects are extremely small and the therefore it can be neglected. Hence the performance of short transmission lines depends upon the resistance and inductance of the line.

Transmission lines having length between 80 km and 200 km, and line voltage between 20kV and 100KV fall into the category called "Medium Transmission lines". Owing to appreciate length and voltage of the line, the charging current is appreciable and therefore the capacitance effect cannot be ignored.

Transmission lines having length above 200km and line voltage above 100 KVfall into the category called "long transmission lines".

It is to be emphasized here that exact solution of any transmission line must be based on the fact that the parameters of any line are not lumped but are distributed uniformly throughout the length of the line. However, results obtained by assuming the constants aslumped for short and medium transmission lines are reasonably accurate.

2.3 Regulation and Efficiency of a Transmission Line

Regulation: When the load is supplied, there is a voltage drop in the line due to resistance and inductance of the line and therefore receiving end voltage V_R is usually less than sending end voltage.

Regulation is defined as the change in voltage at the receiving end when the full load is thrown off, the sending end voltage and supply frequency remain unchanged. It is usually expressed as a percentage of receiving end voltage.

Mathematically it can be expressed as

$$\% \text{ Voltage Regulation} = \frac{V_S - V_R}{V_R} \times 100$$

Knowledge of voltage regulation helps in maintaining the voltage at the load terminals with in prescribed limits by employing suitable voltage control equipment.

Efficiency: When the load is supplied, there are line losses due to ohmicresistance of the line conductors and power delivered at the load end of a transmission line is less than the power supplied at the sending end. It is defined as " Ratio of powerdelivered at the receiving end to the power sent from the sending end. Mathematically it can be expressed as,

$$\%\eta = \frac{V_R I_R \cos\phi_R}{V_3 I_3 \cos\phi_s} \times 100$$

2.4 Short Transmission Lines

In short transmission line, the shunt conductance and shunt capacitance are neglected and so only the series resistance and inductive reactance are to be considered. Due to smaller distance and lower line voltage, the capacitance effects are extremely small and therefore can be neglected. Hence the performance of short transmission lines depends upon the resistance and inductance of the line. Though in an actual line, the resistance and inductance are assumed to be lumped at one place. The equivalent circuit of short transmission line is shown in Fig.

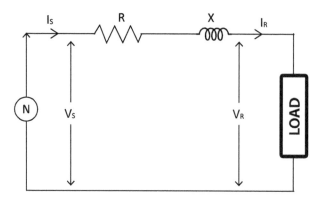

Fig. 2.1 Equivalent Circuit of Short Transmission Line

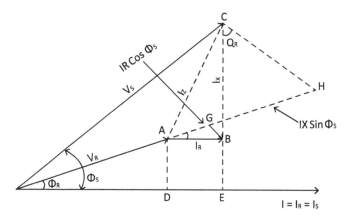

Fig. 2.2 Phasor diagram of Short transmission Line

In a short transmission line, the shunt conductance and shunt capacitance are neglected and so only series resistance and inductive reactance are to be considered. Where R and X represents the total resistance and inductive reactance of the line. Since the effect of stunt capacitance is neglected, the series resistance and reactance are taken as lumped. the equivalent circuit shown in fig represents a 3ϕ short transmission line.

From the equivalent circuit, it is clear that, receiving and voltage,

$$V_R = V_S - I(V + jx)$$

$$= V_S - IZ \text{ (It is the voltage drop along the line)}$$

Sending end voltage $V_S = OC = \sqrt{OE^2 + EC^2}$

$$= \sqrt{(OD+DE)^2 + (EB+BC)^2}$$

$$= \sqrt{(V_R \cos\phi_R + IR)^2 + (V_R \sin\phi_R + I_X)^2}$$

Sending end phase angle $\phi_s = \tan^{-1} \dfrac{EC}{OE}$

$$= \tan^{-1} \dfrac{V_R \sin\phi_R + I_X}{V_R \cos/\phi_R + I_R}$$

Sending end power factor $\cos\phi_s = \dfrac{OE}{OC}$

$$= \dfrac{V_S \cos\phi_R + I_R}{V_S}$$

% Regulation $= \dfrac{V_S - V_R}{V_R} \times 100$

$$= \dfrac{\sqrt{(V_R \cos\phi_R + I_R)^2 + (V_R \sin\phi_R + I_X)^2} - V_R}{V_R} \times 100$$

For determination of an approximate value of sending end voltage V_S may be taken equal to its component along V_R. So, sending end voltage an be written as,

$$V_S \cong OA + AG + GH$$

$$V_S \cong V_R + IR\cos\phi_R + I_X \sin\phi_R$$

% voltage Regulation $\cong \dfrac{V_S - V_R}{V_R} \times 100$

% voltage Regulation $\cong \dfrac{V_R + IR\cos\phi_R + I_X \sin\phi_R - V_R}{V_R} \times 100$

From the above expression it is obvious that voltage regulation of the line depends up on the resistance and reactance of the line.

Regulation per unit $= \dfrac{I_R \cos\phi_R}{V_R} + \dfrac{(I_X \sin\phi_R)}{V_R}$

Regulation per unit $= V_R \cos\phi_R + V_X I \cos\phi_R$

Power delivered to the load $= V_R I \cos\phi_R$

Line losses $= I^2 R$

Power sent from supply end $= V_R I \cos\phi_R + I^2 R$

Efficiency of transmission line $\eta_T = \dfrac{\text{power delivered to the load}}{\text{power supplies from supply}}$

$$\eta_T = \dfrac{V_R \cdot I \cos\phi_R}{V_R \cdot I \cdot \cos\phi_R + I^2 R} \times 100$$

Effect of load P. f on Regulation & Efficiency:

% Regulation $= \dfrac{I_R \cos\phi_R + I X_L \sin\phi_R}{V_R} \times 100$ for lagging P.f

$= \dfrac{I_R \cos\phi_R - I X_L \sin\phi_R}{V_R} \times 100$ for leading P.f

- When load P.f is lagging or unity, the voltage regulation. i.e., $V_R < V_S$
- For a given $V_R \& I$, voltage regulation of the line increases with decrease in P.F for lagging loads.
- When the load P.f is leading, then the voltage regulation is $-ve$ i.e., $V_R > V_S$
- For a given $V_R \& I$, voltage regulation of the line decreases with decrease in P.f for leading loads.

2.5 Generalized Circuit Parameters

In any passive, bilateral and linear network with two input and two output terminals. The input voltage and current can be expressed in terms of output voltage and current. A transmission line is a 4 terminal network. Two input terminals where the power enters and two output terminals where power leaves the network. Such circuit is passive as it does not contain any source of emf. Linear as its impedance is independent of current flowing and bilateral as its impedance is independent of direction of current flowing.

The input voltage and current can be expressed in terms of output voltage and current as,

$$V_S = AV_R + BI_R \quad \rightarrow (1)$$

$$I_S = CV_R + DI_R \quad \rightarrow (2)$$

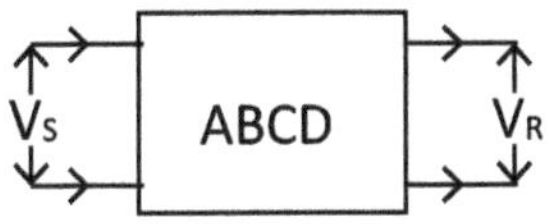

Fig. 2.3 Two port Network of Transmission system

A B C D are called transmission line constants. Once the values of these constants known for a particular transmission line, the performance of the line can be obtained easily.

The value of A, B, C and D can be obtained as follows,

i. with receiving end open-circuited, i.e., $I_R = 0$ to equation (1) & (2) becomes,

$$V_S = AV_R \Rightarrow A = \frac{V_S}{V_R} \text{ which is dimensionless}$$

$$I_S = CV_R \Rightarrow C = \frac{I_S}{V_R}$$

ii. with receiving end S.C., i.e., $V_R = 0$

$$V_S = BV_R \Rightarrow B = \frac{V_S}{V_R} \text{ unit is '}\mho\text{'.}$$

$$I_S = DV_R \Rightarrow D = \frac{I_S}{V_R} \text{ dimensionless}$$

For positive network, $AD - BC = 1$

For symmetrical network, $A = D$

For short transmission lines:

$$V_S = V_R + I_R Z \quad \rightarrow (3)$$

$$I_S = I_R \quad \rightarrow (4)$$

Compare equation (3) with equation (1), then

$$A = 1, \ B = Z$$

Comparing equation (4) with equation (2), then

$$C = 0, D = 1$$

So here,

$A = D \to$ It is symmetrical N/W

And

$AD - BC = 1$ It is passive N/W

2.6 Medium Transmission Lines

For transmission lines of length up to 80km and transmitting power at relatively low voltage (<20KV), the capacitance is too small that its effect can be neglected. However, the effects of shunt capacitance become more and more pronounced with the increase in length and operating voltage of the line. Since medium transmission lines have sufficient length exceeding 80km and usually operate at voltages exceeding 20 kV, the capacitive current is appreciable and therefore line capacitance is to be taken into account. The capacitive current is always flowing in the line while the supply end switches are closed even though the receiving end of the line may be open circuited. The magnitude of capacitive current flowing at any point along the line is that required to charge the section of the line between the given points and receiving end lines it has a max. value at the sending end and diminishes at a practically uniform rate down to zero at the receiving end.

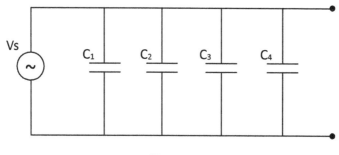

Fig. 2.4

Actually the capacitance of the line is uniformly distributed over its entire length as shown in Fig.2.4. However, in order to make the calculations simple, the capacitance of the system is assumed to be divided and lumped in the form of capacitors shunted across the line at one or more points. The more are the points, the closes the approximation.

The most common methods of representing the medium transmission lines are,

  i. Nominal – T method (Middle condenser Method)
  ii. Nominal – II method (Split condenser Method)

2.6.1 Nominal – T Method:

In this method the whole of the line capacitance is assumed to be concentrated at the middle point of the line and half the line resistance and reactance (i.e., $R/2$ & $X/2$) to be lumped on either side as shown in Fig.

Voltage at the sending end is V_S, at the capacitor is V_C and at the receiving end is V_R. A capacitive current (a) charging current leading the voltage V_C across the capacitor by $90°$. The current in the receiving end half of the line is 'I_R' and in the sending end half of the line is 'I_S'. Which is the phases sum of the receiving end current I_R and charging current I_C.

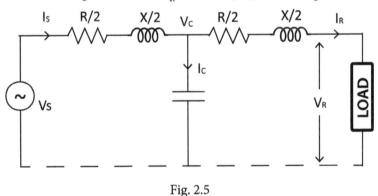

Fig. 2.5

Taking receiving end phase voltage as the reference phasor, we have 'V_R'.

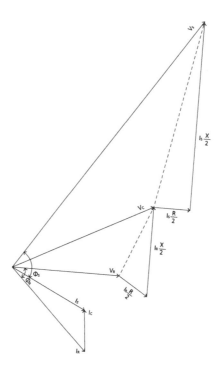

Fig. 2.6 Phasor diagram for Nominal – T method

Receiving end current $I_R = I_R \cos\phi_R - jI_R \sin\phi_R$

$$= I_R (\cos\phi_R - j\sin\phi_R)$$

Voltage across the capacitor $V_C = V_R + I_R \left(\dfrac{R}{2} + j\dfrac{X}{2}\right)$

$$\therefore V_C = V_R + (I_R \cos\phi_R - j I_R \sin\phi_R)\left(\dfrac{R}{2} + j\dfrac{X}{2}\right)$$

Sending end current $I_C = I_R + I_C$

$$I_S = I_R \cos\phi_R - j\sin d_R) + 2\pi f C V_C$$

Sending end voltage $V_S = V_C + I_S \left(\dfrac{R}{2} + j\dfrac{X}{2}\right)$

To calculate regulation, it is required to calculate receiving end no load voltage (V_{R_0}) keeping V_S is fixed in magnitude. The Nominal-T circuit for this condition reduces to the following.

$$V_{R_0} = \frac{X_C}{\frac{R}{2} + j\frac{X}{2} + X_C} \cdot V_S$$

$$= V_S \cdot \frac{\left(\frac{1}{jW_C}\right)}{\frac{R}{2} + j\frac{X}{2} + \frac{1}{jW_C}}$$

$$= V_S \cdot \frac{\left(\frac{-j}{W_C}\right)}{\frac{R}{2} + j\frac{X}{2} - \frac{j}{W_C}}$$

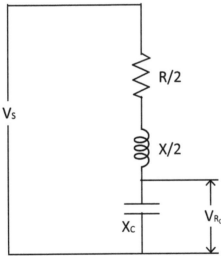

Fig. 2.7 Equivalent circuit at no-load

Now, the voltage regulation for nominal – T can be obtained as

$$\% \text{ Regulation} = \frac{V_{R_0} - V_R}{V_R} \times 100$$

Efficiency can be obtained as,

$$\%\eta = \frac{\text{power delivered to the load}}{\text{power delivered to the load + losses}}$$

$$= \frac{P}{P + I_R^2 \cdot \frac{R}{2} + I_S^2 \frac{R}{2}}$$

$$\%\eta = \frac{P}{P + \frac{R}{2}\left(I_R^2 + I_S^2\right)} \times 100$$

Where 'R' is the resistance per phase. For 3ϕ, the losses are

$$3 \cdot \frac{R}{2}\left(I_R^2 + I_S^2\right)$$

Calculation of ABCD constants for Nominal – T Method:

As we know that,

$$V_S = AV_R + BI_R \quad \rightarrow (1)$$

$$I_S = CV_R + DI_R \quad \rightarrow (2)$$

From the equivalent it diagram,

$$V_S = V_C + I_S \cdot \frac{Z}{2}$$

$$I_S = I_R + I_C$$

Now, $V_C = V_R + I_R \cdot \frac{Z}{2}$

$$I_C = V_C \cdot Y$$

$$\therefore I_S = I_R + I_C$$

$$= I_R + V_C \cdot Y$$

$Z = R + jW_L$

$Y = G + jW_C$

$Y = G + jB$

$Y = 0 + jB$

$Y = G + jB$

$G = 0$

$Y = jB$

$$I_C = \frac{V_C}{X_C} = \frac{V_C}{Y_{jwc}} = \frac{V_C}{\frac{1}{jB}}$$

$$\therefore I_C = \frac{V_C}{\frac{1}{y}} = V_C y$$

$$= I_R + \left(V_R + I_R \cdot \frac{Z}{2}\right) Y \Rightarrow V_R \cdot Y + I_R \left(1 + \frac{YZ}{2}\right)$$

$$V_S = V_C + I_S \cdot \frac{Z}{2}$$

$$= \left(V_R + I_R \cdot \frac{Z}{2}\right) + \left(I_R + \left(V_R + I_R \cdot \frac{Z}{2}\right) Y\right) \frac{Z}{2}$$

$$= \left(V_R + I_R \cdot \frac{Z}{2}\right) + \left(I_R + \left(V_R + I_R \cdot \frac{Z}{2}\right) Y\right) \frac{Z}{2}$$

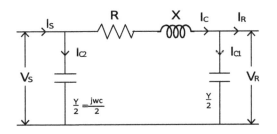

$$= V_R + I_R \cdot \frac{Z}{2} + I_R^2 + V_R \frac{YZ}{2} + I_R \frac{YZ^2}{4}$$

$$V_S = V_{R\left(1+\frac{YZ}{2}\right)} + I_R^2 \left(1 + \frac{YZ}{4}\right) \qquad \rightarrow (3)$$

$$I_S = V_R \cdot Y + I_R \left(1 + \frac{YZ}{2}\right) \qquad \rightarrow (4)$$

Compare equation (3) with equation (1),

$$A = 1 + \frac{YZ}{2}, \quad B = Z\left(1 + \frac{YZ}{4}\right)$$

Compare equation (4) with equation (2)

$$C = Y, \quad D = \left(1 + \frac{YZ}{2}\right)$$

From the above, it is clear that,

$$A = D \text{ symmetrical N/W}$$

$$AD - BC = \left(1 + \frac{YZ}{2}\right)^2 - \left(1 + \frac{YZ}{4}\right).YZ$$

$$AD - BC = 1 \text{ Passive N/W}$$

2.6.2 Nominal – Π Method:

In this method, the capacitance of each line conductor is assumed to be divided into two halves. One half being shunted between line conductor and neutral at the receiving end and the other half at the sending end. The current flowing in the line at any point in between the capacitors is I_C which phaser sum of the load current is I_R and the current drawn by the receiving end capacitor I_{C_1}. Sending end current I_S is the phasor sum of line current I_l and current drawn by the capacitor at the sending end I_{C_2}

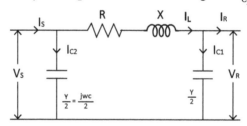

Fig. 2.8 Equivalent circuit of Nominal – Π configuration

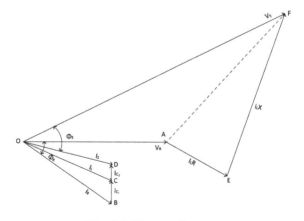

Fig. 2.9 Phasor diagram

Taking V_R as reference phasor,

Sending end voltage, $V_S = V_R + I_l Z$

Sending end current, $I_S = I_l + I_{C_2}$

Charging current at receiving end $I_{C_1} = j\dfrac{W_C}{2}.V_R$

Charging current at sending end $I_{C_2} = j\dfrac{W_C}{2}.V_S$

Line current $I_l = I_R + I_{C_1}$

$$= I_R(\cos\phi_R - j\sin\phi_R) + j\dfrac{W_C}{2}.V_R$$

Sending end voltage $V_S = V_R + \left[I_R(\cos\phi_R - j\sin\phi_R) + j\dfrac{W_C}{2}.V_R\right]Z$

$$V_S = V_R + \left[I_R(\cos\phi_R - j\sin\phi_R) + j\dfrac{W_C}{2}.V_R\right](R+jX)$$

Sending end current $I_S = I_l + I_{C_2}$

$$I_S = \left[I_R(\cos\phi_R - j\sin\phi_R) + j\dfrac{W_C}{2}.V_R\right] + j\dfrac{W_C}{2}.V_S$$

$$V_S = V_R + \left\{I_R(\cos\phi_R - j\sin\phi_R) + j\dfrac{W_C}{2}.V_R\right\}(R+jX)$$

For calculating the regulation of the line, need to find no-load receiving end voltage, thus circuit reduces to

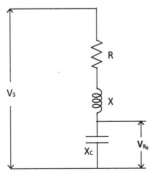

Fig. 2.10 Equivalent circuit at no-load

$$V_{R_0} = V_S \cdot \frac{X_C}{R + jX + X_C}$$

$$= V_S \cdot \frac{j\dfrac{1}{\dfrac{W_C}{2}}}{R + jX + \dfrac{1}{j\dfrac{W_C}{2}}}$$

$$V_{R_0} = V_S \cdot \frac{\left(\dfrac{-2j}{W_C}\right)}{R + jX - \dfrac{2j}{W_C}}$$

$$\therefore \text{Regulation} = \frac{V_{R_0} - V_R}{V_R} \times 100$$

$$\%\eta \text{ (Efficiency)} = \frac{P}{P + I_l^2 R} \times 100$$

$$\%\eta = \frac{P}{P + I_l^2 R} \times 100$$

Where power 'P' is the 3ϕ power delivered at the receiving end and resistance 'R' is the per phase value.

Calculation of A B C D constants for Nominal – II:

As we know that,

$$V_S = AV_R + BI_R \qquad \rightarrow (1)$$

$$I_S = CV_R + DI_R \qquad \rightarrow (2)$$

From equivalent circuit diagram,

$$V_S = V_R + I_l \cdot Z$$

$$I_l = I_R + I_{C_1} \text{ where } I_{C_1} = V_R \cdot \frac{Y}{2}$$

$$\therefore V_S = V_R + \left(I_R + I_R \cdot \frac{Y}{2}\right) Z$$

$$V_S = V_R + I_R \cdot Z + V_R \cdot \frac{YZ}{2}$$

$$V_S = V_R \left(1 + \frac{YZ}{2}\right) + I_R \cdot Z$$

$$V_S = V_R \left(1 + \frac{YZ}{2}\right) + Z \cdot I_R \quad \rightarrow (3)$$

$$I_S = I_1 + I_{C_2}$$

$$= I_R + V_R \cdot \frac{Y}{2} + V_S \cdot \frac{Y}{2}$$

$$= I_R + V_R \cdot \frac{Y}{2} + \left(\left(1 + \frac{YZ}{2}\right)V_R + I_R \cdot Z\right)\frac{Y}{2}$$

$$= I_R + V_R \cdot \frac{Y}{2} + \frac{Y}{2}V_R + \frac{Y^2 Z}{4} \cdot V_R + \frac{YZ}{2} \cdot I_R$$

$$= V_R \left(Y + \frac{Y^2 Z}{4}\right) + I_R \left(1 + \frac{YZ}{2}\right)$$

$$I_S = V_R Y \left(1 + \frac{YZ}{4}\right) + I_R \left(1 + \frac{YZ}{2}\right) \quad \rightarrow (4)$$

Compare equation (3) with Eq. (1),

$$A = Y \left(1 + \frac{YZ}{4}\right) \quad D = 1 + \frac{YZ}{2}$$

So, $A = D$ symmetrical N/W

and $AD - BC = \left(1 + \frac{YZ}{2}\right)^2 - YZ\left(1 + \frac{YZ}{4}\right)$

$AD - BC = 1$ passive N/W

1. *An overhead 3φ line delivers 5MW at 22KV at 0.8 lagging P.f the resistance and reactance of each conductor is 4Ω and 6Ω respectively. Find*
   (i) Sending end voltage (V_s)
   (ii) % Regulation
   (iii) Total line losses
   (iv) Transmission efficiency (η)

Sol: Phase voltage at receiving end $V_R = \dfrac{22 \times 10^8}{\sqrt{3}} = 12,702\,V$

Power delivered $P = 5\,MW = 5 \times 10^6\,W$

Load P.F $\cos\phi_R = 0.8\,lag$

$\sin\phi_R = 0.6$

Line current $I_l = \dfrac{P}{\sqrt{3}V_L \cos\phi}$

$= \dfrac{5 \times 10^6}{\sqrt{3} \times 22 \times 10^3 \times 0.8}$

$= 164\,A$

Sending end voltage / phase $V_S = V_R + I.R\cos\phi_R + I.V.\sin\phi_R$

$= 12702 + 164 \times 4 \times 0.8 + 164 \times 6 \times 0.6$

$= 13,817.2\,volts$

Sending end line voltage $V_{SL} = \sqrt{3} \times 13817.2$

$= 23.932\,V$ (or) $23.93\,KV$

% Regulation $= \dfrac{V_S - V_R}{V_R} \times 100$

$= \dfrac{23932 - 22000}{22000} \times 100$

$= 8.78\%$

Total losses $= 3I^2 R = 3(164)^2 \times 4$

$= 322.752\,KW$

Transmission efficiency $\eta = \dfrac{5\times 10^6}{5\times 10^6 + 322.752\times 10^3}\times 100$

$= 93.93\%$

2. **A 15km long 3φ overhead line delivers 5MW at 11KV at 0.8 lagging P.f line loss is 12% of power delivered line inductance is 1.1 MH/Km/ph. Find sending and voltage and regulation.**

Sol: Power delivered $P = 5MW$

Phase voltage at receiving end $V_R = \dfrac{11\times 10^3}{\sqrt{3}} = 6351V$

Line current $I_l = \dfrac{5\times 10^6}{\sqrt{3}\times 11\times 10^3 \times 0.8} = 3284$

Line loss $= 12\%$ of power delivered

$3I^2R = \dfrac{12}{10}\times 5\times 10^6$

$3I^2R = 600\,KW$

$R = \dfrac{600\times 10^3}{3\times (328)^2}$

$R = 1.859\,\Omega/Ph$

Inductive reactance/Ph $X_L = 2\pi f L$

$= 2\pi \times 50 \times 1.1 \times 10^{-3}$

$= 5.1836\,\Omega/Ph$

Phase voltage at the sending end $V_S = V_R + IR\cos\phi_R + I\sin\phi_R$

$V_S = 6351 + 328\times 1.859\times 0.8 + 328\times 5.126\times 0.6$

$V \quad 7858.93V$

Line voltage at sending end $V_\Omega = \sqrt{3}V_S$

$= \sqrt{3}\times 7858.93$

$= 13612 V$ (or) $13.612 KV$

% Regulation $= \dfrac{V_S - V_R}{V_R} \times 100$

$= \dfrac{13.612 - 11}{11} \times 100$

$= 23.746\%$

3. **A 3ϕ, 50Hz transmission line has resistance, inductance and capacitance per phase of 10Ω, 0.1H and 0.9μf respectively and delivers a load of 35MW at 132KV and 0.8 P.f lagging. Find the efficiency and regulation of the line using nominal – II method.**

Sol: The resistance of the line/phase $R = 10\Omega$

Inductance/phase $L = 0.1H$

Capacitance/phase $C = 0.9\,\mu F$

Inductance reactance $X_L = 2\pi f L$

$= 2\pi \times 50 \times 0.1$

$= 31.4\Omega$

Impedance $Z = R + jX_L$

$= (10 + j31.4)\,\Omega$

Taking V_R as reference,

Receiving end current $I_R = I_R(\cos\phi_R - j\sin\phi_R)$

$I_R = 191.35(0.8 - j0.6)$

$= 153.08 - j\,114.81$

Charging current at receiving end $I_{C_1} = \dfrac{jW_C}{2}.V_R$

$= j2\pi \times 50 \times \dfrac{0.9}{2} \times 10^{-6} \times 76201.2$

$= j\,10.774$

Line current $I_l = I_R + I_{C_1}$

$$= 153.08 - j114.81 + j\,10.77$$

$$= 153.08 - j\,104.04$$

Sending end voltage $V_S = V_R + I_l \cdot Z$

$$= V_R + I_l(R + jX_L)$$

$$= 76210.2 + (153.08 - j\,104.04)(10 + j\,31.)$$

$$= 76210.2 + 153.08 + j\,4606.7 - j\,1040.4 + 3266.85$$

$$= 81007.85 + j\,3766.3$$

$$= 81095.3 \angle 2.66$$

$$|V_S| = 81095.35\,V$$

No-load receiving end voltage

$$V_{R_0} = V_S \cdot \frac{\left(\frac{-2j}{W_C}\right)}{R + jX - \frac{2j}{W_C}} = V_S \cdot \frac{(-j7073)}{10 + j31.4 - j703.5} = 81456.86$$

$$\%\text{ Regulation} = \frac{81456.86 - 76210.2}{76210.2} \times 100$$

$$= 6.88\%$$

Line current $I_l = 153.08 - j104.08$

$$= 185.08 \angle -34.2$$

Losses $= 3I_l^2 R$

$$= 3(185.08)^2 \times 10$$

$$= 1.027\,MW$$

$$\%\text{ Efficiency} = \frac{35}{35 + 1.027} \times 100$$

$$\eta = 97.1\%$$

4. **A single phase OHTL is delivering 600KVA load at 2KV. It's resistance and reactance are 0.18Ω and 0.36Ω respectively. Find the voltage regulation if the load power factors is (i) 0.8 lagging and (ii) 0.8 leading.**

Sol: Given data:

Power transmitted $P = 600\,KVA$

Receiving end voltage $V_R = 2KV = 2000\,V$

Resistance of the line $R = 0.18\,\Omega$

Reactance of the line $X = 0.36\,\Omega$

Load power factor $\cos\phi_R = 0.8$ /lag & lead

Line current $I_l = \dfrac{600 \times 10^3}{2000} = 300\,A$

(i) When $\cos\phi_R = 0.8$ lagging, then $\sin\phi_R = 0.6$

$$V_S = \sqrt{(V_R \cos\phi_R + IR)^2 + (V_R \sin\phi_R + IX)^2}$$

$$= \sqrt{(2000 \times 0.8 + 300 + 0.18)^2 + (2000 \times 0.6 + 300 \times 0.36)^2}$$

$$= 2108.7\,volts$$

$$\therefore \text{voltage regulation} = \dfrac{V_S - V_R}{V_R} \times 100$$

$$= \dfrac{2108.7 - 2000}{2000} \times 100$$

$$= 5.43\%$$

(ii) when $\cos\phi_R = 0.8$ leading

$$V_S = \sqrt{(V_R \cos\phi_R + IR)^2 + (V_R \sin\phi_R + IX)^2}$$

$$\cong 1982\,V$$

$$\%\text{ Regulation} = \dfrac{V_S - V_R}{V_R} \times 100$$

$$= \dfrac{1982 - 2000}{2000} \times 100 = 0.9\%$$

5. An overhead 3ϕTransmission line delivers 4000kw at 11KV at 0.8 power factor lagging. The resistance and reactance of each conductor are 1.5Ω and 4Ω/phase respectively. Find
   (i) The line sending end voltage
   (ii) % Regulation
   (iii) Transmission efficiency

6. A balanced 3ϕ load of 30MW is supplied at 132KV, 50Hz and 0.85Pf lagging by means of a transmission line. The series impedance of a simple conductor is $(20+j52)\Omega$ and total phase neutral admittance is $315\times10^{-6}\mho$. Using nominal – I method, find
   (i) The A B C D constants of the line
   (ii) V_S
   (iii) % Regulation

Sol: Given that, Impedance $Z = (20+j52)\Omega$

Admittance $Y = j\ 315\times10^{-6}\ \mho$

Receiving end voltage $V_R = 132\,KV$ (line), $\cos\phi_R = 0.85$

Power delivered to the load $P = 30MW$, frequency $f = 50Hz$

7. Using Nominal – T method calculate A B C D constants of 3-ϕ, 80km, 50Hz transmission line with series impedance of $(0.15+j\,0.78)\Omega/km$ and shunt admittance of $j(5\times10^{-6})\mho/km$

Sol: Given data:

Length of transmission line $= 80km$

Series impedance per km $\bar{Z} = (0.15+j0.78)\ \Omega$

Shunt admittance per km $\bar{Y} = j(5\times10^{-6})\mho$

Impedance of 80km line $\bar{Z} = (0.15+j0.78)\times 80$
$$= (12+j\,62.4)\ \Omega$$

Total admittance of line $\bar{Y} = j(5\times10^{-6})\times 80$
$$= j4\times10^{-4}\ \mho$$

$$A = 1 + \frac{YZ}{2}$$

$$= 1 + \frac{(j4\times 10^{-4})(12 + j\ 62.4)}{2} = (0.987 + j\ 0.002)$$

$$B = Z\left(1 + \frac{YZ}{4}\right)$$

$$= (12 + j62.4)\left(1 + \frac{(j4\times 10^{-4})(12 + j\ 62.4)}{4}\right)$$

$$= (11.85 + j62.02)$$

$$C = Y = j4\times 10^{-4}\ \mho$$

$$D = A = (0.987 + j\ 0.002)$$

8. **A 3-φ overhead transmission line delivers a load of 80MW at 0.8 P.f lag and 220KV between the lines. Its total series impedance and shunt admittance per phase are $200\angle 80°\ \Omega$ and $0.0013\angle 90°\ \Omega$ respectively. Using Nominal T and nominal - Π method calculate.**
   (i) A, B, C, D constants of line
   (ii) sending end voltage
   (iii) sending end current
   (iv) sending end P.f
   (v) Transmission efficiency
   (vi) % Regulation

Sol: Given data:

Power delivered $= 80MW$

Receiving end voltage $V_{R_{line}} = 220KV$

$$V_{R_{Ph}} = \frac{220}{\sqrt{3}} = 127.01 KV\ /\ Ph$$

Impedance $Z = 200\angle 80°\ \Omega\ /\ Ph$

Shunt admittance $Y = 0.0013\angle 90°\ \mho\ /\ Ph$

$$I_{R_{Ph}} = \frac{P}{\sqrt{3} \times V_R \times \cos\phi_R}$$

$$= \frac{80 \times 10^6}{\sqrt{3} \times 220 \times 10^3 \times 0.8} = 262.431 A$$

Take V_R reference phasor

$$I_R = I(\cos\phi_r - j\sin\phi_r)$$

$$= 262.43(0.8 - j0.6)$$

$$= (209.94 - j157.4)$$

$$= 262.43 \angle -36.86° A$$

1. **Nominal – T method:**

(i) A, B, C, D constants of line

$$A = 1 + \frac{YZ}{2} = 1 + \frac{(0.0013 \angle 90°)(200 \angle 80°)}{2} = 0.871 \angle 1.48°$$

$$B = Z\left(1 + \frac{YZ}{4}\right) = 200 \angle 80° \left(1 + \frac{(0.0013 \angle 90°)(200 \angle 80°)}{4}\right)$$

$$= 187.21 \angle 80.6$$

$$C = Y = 0.0013 \angle 90°$$

$$D = A = 0.871 \angle 1.48°$$

(ii) Sending end voltage:

$$V_S = AV_R + BI_R$$

$$= (0.871 \angle 1.48)(127.01 \times 10^3) + (187.21 \angle 80.6°)(262.43 \angle -36.86)$$

$$= (110588.8 + j\, 2857.2) + (35495.32 + j33967.51)$$

$$= 150.65 \angle 14.14° \, KV/Ph$$

(iii) Sending end current:

$$I_S = CV_R + DI_R$$

$$= (0.0013\angle 90°)(127.01\times 10^3) + (0.871\angle 1.48)(262.43\angle -36.86)$$

$$= 189.22\angle 9.97° \, A$$

(iv) **Sending end power factor:**

$$\cos\phi_s = \cos(14.14 - 9.97)$$

$$= 0.9973$$

(v) **Transmission efficiency:**

$$\eta = \frac{output\ power}{input\ power} \times 100$$

$$= \frac{80\times 10^6}{\sqrt{3}\times(260.93\angle 14.14°)\times 10^3 \times 189.22\angle 9.97 \times 0.9973} \times 100$$

$$= 93.80\ \%$$

(vi) **% Regulation:**

We know $V_S = AV_R + BI_R$

At no load $I_R = 0$

$$V_S = AV_{R_0}$$

$$V_{R_0} = \frac{V_S}{A} = \frac{150.65\angle 14.14}{0.871\angle 1.48} = 172.96\angle 12.66\ KV$$

$$\%\ Regulation = \frac{V_{R_0} - V_R}{V_R} \times 100$$

$$= \frac{172.96 - 127.01}{127.01} \times 100 = 36.14\%$$

2. **Nominal - II method:**

(i) $A = 1 + \dfrac{YZ}{2} = 1 + \dfrac{(0.0013\angle 90°)(200\angle 80°)}{2} = 0.872\angle 1.48$

$B = Z = 200 \angle 80$

$C = Y\left[1+\dfrac{YZ}{4}\right] = (0.0013 \angle 19)\left[1+\dfrac{(0.0013 \angle 90)(200 \angle 80)}{4}\right]$

$= 1.216 \times 10^{-3} \angle 90.69$

$D = A = 0.872 \angle 1.48$

(ii) Sending end voltage:

$$V_S = AV_R + BI_R$$

$$= (0.872 \angle 1.48)(127.01 \times 10^3) + (200 \angle 80)(262.43 \angle -36.86)$$

$$= 153.969 \angle 14.5 \; KV/Ph$$

(iii) Sending end current:

$$I_S = CV_R + DI_R$$

$$= (1.216 \times 10^{-3} \angle 91.48)(127.01 \times 10^3) + (0.872 \angle 1.48)(262.43 \angle -36.86)$$

$$= 154.44 \angle 91.48 + 228.83 \angle -35.38$$

$$= 186.20 \angle 6.77° \; A$$

(iv) Sending end power factor:

$$\cos \phi_S = \cos(14.5 - 6.77) = 0.99$$

(v) Transmission efficiency:

$$\eta = \dfrac{output \; power}{imput \; power} \times 100$$

$$= \dfrac{P}{\sqrt{3} \times V_S \times I_S \times \cos \phi_S} \times 100$$

$$= \dfrac{80 \times 10^6}{\sqrt{3} \times 266.68 \times 186.01 \times 0.99 \times 10^3} \times 100 = 94.05\%$$

(vi) % Regulation:

$$V_{R_0} = \frac{V_S}{A} = \frac{153.96 \angle 14.5}{0.872 \angle 1.48} = 176.55 \angle 13.02\,KV$$

$$\% \text{ Regulation} = \frac{V_{R_0} - V_R}{V_R} \times 100 = \frac{176.55 - 127.01}{127.01} \times 100 = 39\%$$

Problems on short transmission lines:

9. *An overhead 3-φ line delivers 5MW at 22KV at 0.8 lagging the resistance and reactance of each conductor is 4Ω and 6Ω respectively. Find (i) sending end power factor (ii) percentage regulation (iii) total line losses (iv) transmission efficiency?*

Sol: Given data:

Power delivered, $P = 5MW$

Receiving end voltage, $V_R = 22KV$

Power factor, $\cos\phi_R = 0.8$ lagging $\Rightarrow \sin\phi_R = \sin(\cos^{-1} 0.8))$

$\qquad\qquad\qquad\qquad\qquad\qquad\qquad\qquad = 0.6$

Resistance, $R = 4\Omega$

Reactance, $X = 7\Omega$

Receiving end current $I_R = \dfrac{\text{Power delivered}}{\sqrt{3} \times V_R \times \cos\phi_R} = \dfrac{5 \times 10^6}{\sqrt{3} \times 22 \times 10^3 \times 0.8}$

$\qquad\qquad\qquad\qquad I_R = 164.02\,Amps$

(i) Sending end voltage

$V_S = V_R + IR\cos\phi_R + I_X \sin\phi_R$

$\quad = 22 \times 10^3 + (164.02 \times 0.8 \times 4) + (164.02 \times 0.6 \times 6)$

$V_S = 23115.33\,V$

(ii) Percentage regulation

$= \dfrac{V_S - V_R}{V_R} \times 100 = \dfrac{23115.33 - 2 \times 10^3}{22 \times 10^3} \times 100$

% V.R $= 5.069\,\%$

(iii) Total line losses $= 3I^2R$

$$= 3 \times (164.02)^2 \times 4$$

$$= 322.830 \, KW$$

(iv) Transmission efficiency, $\eta = \dfrac{Power\ delivered}{Power\ delivered + losses} \times 100$

$$\eta = \dfrac{5 \times 10^6}{5 \times 10^6 \times 322.83 \times 10^3} \times 100 = 93.9\%$$

10. **A 3-ϕ overhead transmission line delivers 5MW at 11KV at 0.8 lagging power factor. The line losses are 12% of power delivered, line inductance is 1.1MH. Find the sending end voltage, regulation and efficiency?**

Sol: *Given data:*

Power delivered, $P = 5MW$

Receiving end voltage, $V_R = 11KV$

Receiving end power factor, $\cos\phi_R = 0.8$ lagging $\Rightarrow \sin\phi_R = 0.6$

Line losses $= 12\%$ of power delivered

$$3I^2R = 0.12 \times 5 \times 10^6$$

$$= 0.6 \, MW$$

Line inductance, $L = 1.1MH$

Receiving end current, $I_R = \dfrac{Power\ delivered}{\sqrt{3} \times V_R \times \cos\phi_R} = \dfrac{5 \times 10^6}{\sqrt{3} \times 11 \times 10^3 \times 0.8}$

$$I_R = 328.03 \, Amps$$

$$3(328.03)^2 \times R = 0.6 \times 10^6$$

$$R = \dfrac{0.6 \times 10^6}{3 \times (328.03)^2} = 1.25 \, \Omega$$

$$X_L = 2\pi f l = 2 \times \pi \times 50 \times 1.1 \times 10^{-3} = 0.345 \, \Omega$$

(i) Sending end voltage $V_S = V_R + IR\cos\phi_R + IX\sin\phi_R$

$$= 11\times10^3 + 328.03\times1.25\times0.8 + 328.03\times0.345$$

$$V_S = 11553.38 V$$

(ii) Percentage regulation

$$= \frac{V_S - V_R}{V_R} \times 100 = \frac{11553.38 - 11\times10^3}{11\times10^3} \times 100$$

$\%VR = 5.03\%$

(iii) Efficiency $\eta = \frac{Power\ delivered}{Power\ delivered + losses} \times 100$

$$= \frac{500\times10^6}{5\times10^6 + 0.6\times10^6} \times 100 = 89.28\%$$

11. *A 1-φ over head transmission line delivers 1100KW at 33KΩ at 0.8 power factor lagging. The total resistance and the inductive reactance of the line are 10 Ω and 15 Ω respectively. Determine (i) sending end voltage (ii) sending end power factor (iii) Transmission efficiency?*

Sol: Given data:

Power delivered, $P = 1100\ KW$

Receiving end voltage $V_R = 33 KV$

Receiving end power factor, $\cos\phi_R = 0.8$ lagging $\Rightarrow \sin\phi_R = 0.6$

Resistance, $R = 10\Omega$

Inductance reactance, $X_L = 15\Omega$

(i) Sending end voltage, $V_S = V_R + IR\cos\phi_R + IX\sin\phi_R$

Receiving end current $I_R = \frac{Power\ delivered}{rV_R \times \cos\phi_R} = \frac{1100\times10^3}{\sqrt{3}\times33\times10^3\times0.8}$

$I_R = 41.46\ Amps$

$V_S = 33 \times 10^3 + 41.66 \times 10 \times 0.8 + 41.66 \times 15 \times 0.6$

$V_S = 33708.22\,V$

(ii) Sending end power factor,

$$\cos\phi_S = \frac{V_R \cos\phi_R + IR}{V_S} = \frac{33 \times 10^3 \times 0.8 + 41.66 \times 10}{33708.22}$$

$\cos\phi_S = 0.79$

(iii) Efficiency $\eta = \dfrac{1100 \times 10^3}{1100 \times 10^3 \times 17355.55} \times 100 = 98.44\%$

Losses $I^2 R = (41.66)^2 \times 10 = 17355.55\,W$

12. **A 3-ϕ, 50Hz transmission line has resistance, inductance and capacitance per phase of 10Ω, 0.1 H & 0.9μf respectively and delivers a load of 35 MW at 132 KV at 0.8 power factor lagging. Find, efficiency and regulation of the line using nominal - II method.**

Sol: *Given data:*

Resistance per phase, $R = 10\,\Omega$

Frequency, $f = 50\,Hz$

Inductance per phase, $L = 0.1\,H$

Capacitance per phase, $C = 0.9\,\mu f$

Power delivered $= 35\,MW$

Receiving end voltage, $V_R = 132\,KV$

Receiving end power factor, $\cos\phi_R = 0.8\,P.f$ (lag)

Inductance reactance $X_L = 2\pi f L = 2 \times \pi \times 50 \times 0.1 = 31.4\,\Omega$

Impedance, $Z = R + j X_L = (10 + j31.4)\,\Omega$

Receiving end current, $I_R = \dfrac{\text{Power delivered}}{\sqrt{3} \times V_R \times \cos\phi_R} = \dfrac{35 \times 10^6}{\sqrt{3} \times 132 \times 10^3 \times 0.8}$

$I_R = 191.35$ Amps

$\overline{I_R} = 191.35\,(\cos\phi_R - j\sin\phi_R) = 191.35(0.8 - j0.6)$

$\overline{I_R} = (153.08 - j114.81)$ Amps

$I_{C_1} = jW\dfrac{C}{2}V_R = j\times 2\pi\times 50\times \dfrac{0.9\times 10^{-6}}{2}\times \dfrac{132\times 10^3}{\sqrt{3}}$

$I_{C_1} = j10.77$ Amps

$I_L = I_R + I_{C_1} = (153.08 - j\,114.81) + j10.77$

$ = (153.08 - j104.04)$ Amps

$V_S = V_R + I_L(R + jX)$

$= \dfrac{132\times 10^3}{\sqrt{3}} + (153.08 - j104.04)(10 + j31.4)$

$= 76210.23 + 1530.8 + j4806.71 - j1040.4 + 3266.85$

$V_S = (81007.85 + j3766.3)$ V

No load receiving end voltage $V_{R_0} = V_S\left(\dfrac{\dfrac{-2j}{WC}}{R + jXL - \dfrac{2j}{WC}}\right)$

$= (81007.85 + j3766.3)\left(\dfrac{\dfrac{-2j}{2\pi\times 50\times 0.9\times 10^{-6}}}{10 + j(2\pi\times 50\times 0.1) - \dfrac{2j}{2\times\pi\times 50\times 0.9\times 10^{-6}}}\right)$

$= 81457.06\angle -177.24°$

$V_{R_0} = 81457.06(-0.99 - j0.048)$

$V_{R_0} = 81456.86\,V$

% voltage regulation $= \dfrac{V_{R_0} - V_R}{V_R}\times 100 = \dfrac{81456.86 - 76210.23}{76210.23}\times 100$

%V.R. = 6.8 %

$$\text{Efficiency, } \%\eta = \frac{\text{Power delivered}}{\text{Power delivered} + \text{losses}} \times 100$$

$$\text{Losses} = 3I_L^2 \times R = 3(185.08)^2 \times 10 \qquad [\because |I_L| = 185.08]$$

$$= 1.02 MW$$

$$\text{Efficiency, } \%\eta = \frac{35 \times 10^6}{35 \times 10^6 + 1.02 \times 10^6} \times 100 = 97.16\%$$

Case II: Using nominal – T method:

$$I_R = 191.35 \ (\cos\phi_R - j\sin\phi_R) = (153.08 - j114.81) \ Amps$$

$$V_C = \overline{V_R} + I_R \ (\cos\phi_R - j\sin\phi_R)\left(\frac{R}{2} + j\frac{X}{2}\right)$$

$$= 76210.23 + (153.08 - j114.81)\left(\frac{10}{2} + j\frac{0.1}{2}\right)$$

$$V_C = (76981.37 - j566.4) \ volts$$

$$I_C = jWCV_C$$

$$= j(2\pi \times 50) \times 0.9 \times 10^{-6} \times (76981.37 - j566.4)$$

$$= j \ 2.82 \times 10^{-4} (76981.37 - j566.4)$$

$$I_C = (j21.76 + 0.156) \ Amps$$

$$I_S = I_R + I_C = (153.23 - j114.81) + (0.156 + j21.76)$$

$$I_S = (153.23 - j93.05) \ Amps$$

Sending end voltage,

$$V_S = V_C + I_S\left(\frac{R}{2} + j\frac{X}{2}\right)$$

$$V_S = (776981.37 - j566.4) + (153.23 - j93.05)(5 + j0.05)$$

$$V_S = (77752.17 - j915.16) \ volts$$

Receiving end voltage,

$$V_{R_0} = V_S \left(\frac{\dfrac{-j}{WC}}{\dfrac{R}{2} + j\dfrac{X}{2} - \dfrac{j}{WC}} \right)$$

$$= (77752.17 - j915.16) \left(\dfrac{-j}{5 + j0.05 - \dfrac{j}{2\pi \times 50 \times 0.9 \times 10^{-6}}} \right)$$

$V_{R_0} = (77751.82 - j1025.09)\,volts$

$$\text{Percentage regulation} = \dfrac{V_{R_0} - V_R}{V_R} \times 100 = \dfrac{77715.82 - 76210.23}{76210.23} \times 100$$

$\%V.R = 2.02\%$

$$\text{Efficiency, } \%\eta = \dfrac{\text{Power delivered}}{\text{Power delivered} + \text{losses}} \times 100$$

$$\text{Losses} = 3(I_S^2 + I_R^2)\dfrac{R}{2}$$

$$= 3(179.27^2 + 191.35^2) \times \dfrac{10}{2} = 1031288.33\ W$$

$$\%\eta = \dfrac{35 \times 10^6}{35 \times 10^6 + 1031288.33} \times 100$$

$\%\eta = 97.13\%$

13. **Using nominal – T method, calculate A, B, C, D constants of 3-ϕ, 80km, 50Hz, transmission line with series impedance of $(0.15 + j0.78)\,\Omega/km$ and shunt admittance of $j(5 \times 10^{-6})\,\mho/km$ and shunt admittance of $j(5 \times 10^{-6})\,\mho/km$?**

Sol: *Given data:*

Series impedance, $Z = (0.15 + j0.78)\ \Omega/km$

For 80km, $Z = (0.15 + j0.78) \times 80 = (12 + j62.4)\,\Omega$

Shunt admittance, $Y = j(5 \times 10^{-6})\,\mho/km$

For nominal – T method, A, B, C, d constants are,

$$A = \left(1 + \dfrac{YZ}{2}\right),\ B = Z\left(1 + \dfrac{YZ}{4}\right),\ C = Y,\ D = 1 + \dfrac{YZ}{2}$$

$$A = 1 + \frac{j(4\times10^{-4})(12+j62.4)}{2} = (0.987 + j\,2.4\times10^{-3})$$

$$A = D = (0.987 + j2.4\times10^{-3})$$

$$B = Y\left(1+\frac{YZ}{4}\right) = j(4\times10^{-4})\left(1+\frac{j(4\times10^{-4})(12+j\,62.4)}{4}\right)$$

$$B = (11.85 + j\,62.02)\ \Omega$$

$$C = Y = j(4\times10^{-4})\ \mho$$

14. **A 3-ϕ, 50Hz overhead transmission line 10km long has the following constants. Resistance /km /Ph is 0.1 Ω, inductive reactance /km /Ph is 0.2 Ω and capacitive suceptance /km /Ph is 0.04×10^{-4} siemens. Find (i) sending end current (ii) sending end voltage (iii) sending end power factor (iv) transmission efficiency. When supplying a balanced load of 10,000 KW at 66 KV, 0.8 power factor lagging. Use nominal - I method?**

Sol: Given data:

For 100 km, resistance / Ph $= 0.1\times100 = 10\,\Omega$

For 100 km, reactance /Ph, $X_L = 0.2\times100 = 20\,\Omega$

For 100 km, capacitive susceptance, $Y = 100\times0.04\times10^{-4}$
$$= 4\times10^{-4}\ siemens$$

Power delivered, $P = 10000\ KW$

Receiving end voltage, $V_R = 66\ KV$

$$(V_R)_{Ph} = 38105.117\ V$$

Receiving end current, $I_R = \dfrac{Power\ delivered}{\sqrt{3}\times V_R \times \cos\phi_R}$

$$I_R = \frac{10000\times10^3}{\sqrt{3}\times66\times10^3\times0.8} = 109.34\ Amps$$

Charging current, $I_C = V_C y$

$$V_C = V_R + I_R (\cos\phi_R - j0.6)\left(\frac{Z}{2}\right)$$
$$= 38105.117 + 109.34(0.8 - j0.6)(5 + j10)$$
$$V_C = (39198.517 + j546.7) \text{ volts}$$

$$I_C = V_C j y$$
$$= (39198.517 + j\,546.7) \times (4 \times 10^4)$$
$$I_C = (-0.218 + j15.67) \text{ Amps}$$

(i) Sending end current, $I_S = I_R + I_C$

$$I_S = (87.47 - j\,65.6) + (-0.128 + j\,15.67)$$
$$= (87.34 - j\,49.93) \text{ Amps}$$
$$I_S = 100.6 \angle -29.7 \text{ Amps}$$

(ii) Sending end voltage,

$$V_S = V_C + I_S \left(\frac{R}{2} + j\frac{X}{2}\right)$$
$$= (39198.517 + j546.7)(87.34 - j49.93)(5 + j10)$$
$$V_S = (40134.52 + j1170.47) \text{ volts}$$
$$V_S = 40151.59 \angle 1.67° \text{ volts}$$

(iii) sending end power factor, $\cos\phi_S = \cos(\text{Angle of '}V_s\text{'} - \text{angle of '}I_S\text{'})$
$$= \cos(1.67 - (-29.7))$$
$$= 0.85 \text{ (lag)}$$

Power input $= 3V_S I_S \cos\phi_S$
$$= 3 \times 40151.59 \times 100.6 \times 0.85$$
$$= 10.3 \text{ MW} = 10328 \text{ KW}$$

(iv) Efficiency

$$= \frac{Power\ delivered}{Power\ delivered + losses} \times 100 = \frac{Power\ delivered}{Power\ input} \times 100$$

$$= \frac{10000 \times 10^3}{10328 \times 10^3} \times 100 = 97.1\%$$

Case 2: Using nominal – Π method:

Receiving end current, $I_R = 109.34(0.8 - j0.6)$

$$= (87.47 - j65.60)\ Amps$$

$$I_{C_1} = JW\overline{V_R} = 38105.12 + j\frac{4 \times 10^{-4}}{2}$$

$$I_{C_1} = j\ 7.621\ Amps$$

$$I_L = I_R + I_{C_1} = (87.472 - j65.60) + j\ 7.621$$

$$I_C = (87.2 - j57.77)\ Amps$$

$$I_L = 104.94 \angle -33.53°\ Amps$$

Sending end voltage, $V_S = V_R + I_Z$

$$= 38105 + (104.94 \angle -33.53°)(10 + j\ 20)$$

$$= 40156.3 \angle 1.66°\ V = (40139.28 + j\ 1169.6)\ volts$$

$$\theta_{V_S} = 1.66°$$

Charging current at sending end, $I_{C_2} = jV_S\dfrac{Y}{2}$

$$= j\ 40156.3 \angle 1.66 \times \frac{40 \times 10^{-4}}{2}$$

$$I_{C_2} = (-0.23 + j\ 8.027)\ Amps$$

$$I_{C_2} = 8.03 \angle 91.66°\ Amps$$

Sending end current, $\overline{I_s} = \overline{I_c} + \overline{I_{c_2}}$

$$= (87.47 - j57.979) + (-0.23 + j8.027)$$

$$I_s = (87.23 - j\,49.95)\ Amps$$

$$I_s = 100.52 \angle -29.79°$$

$$\theta_{I_s} = -29.79°$$

Sending end phase angle, $\theta_s = \theta_{V_s} + \theta_{I_s}$

$$= 1.66 - (-29.79)$$

$$\theta_s = 31.45°$$

Sending end power factor, $\cos\phi_s = \cos(31.45) = 0.853$ lag

Line losses $3I^2 R = 3(104.94)^2 \times 10 = 330372.1\ W$

Transmission efficiency, $\eta = \dfrac{Power\ delivered}{Power\ delivered + losses} \times 100$

$$= \dfrac{10000 \times 10^3}{10000 \times 10^3 + 330372.1} \times 100$$

$$\eta = 96.8\%$$

15. **A 3φ, 50 Hz, 150 km line has a resistance, inductive reactance and capacitive shunt admittance of 0.1 Ω, 0.5 Ω and 3×10^{-6} seimen/km/Ph. If the line delivers 50 MW of 120 KV and 0.8 P.f lagging find (i) sending end current (ii) Sending end voltage (iii) sending end power factor, (iv) Regulation (v) efficiency. Assume nominal - II configuration?**

Sol: Given data:

Length of the line, $= 150\ km$

Resistance /km/Ph $= 0.1\ \Omega$

For 150 km, R/Ph $= 0.1 \times 150 = 15\ \Omega$

Inductance reactance /Ph / km $= 0.5\,\Omega$

For 150 km, X_L/Ph $= 0.5 \times 150 = 75\,\Omega$

Capacitive shunt admittance /Ph/km $= Y = 3 \times 10^{-6}\,S$

For 150 km, Y/Ph $= 3 \times 10^6 \times 150 = 4.5 \times 10^{-4}\,s$

Power delivered, $P = 50\,MW$

Receiving end voltage $V_R = 110\,KV$

$$V_R / Ph = \frac{(110 \times 10^3)}{\sqrt{3}} = 63508.52\,V$$

Receiving end power factor, $\cos\phi_R = 0.8$ lag

Receiving end current, $I_R = \dfrac{P_R}{\sqrt{3} \times V_R \times \cos\phi_R} = \dfrac{50 \times 10^6}{\sqrt{3} \times 110 \times 10^3 \times 0.8}$

$$I_R = 328\,Amps$$

$$\cos\phi_R = 0.8,$$
$$\sin\phi_R = 0.6$$

$$\overline{I_R} = I_R(\cos\phi_R + j\sin\phi_R) = 328.03(0.8 - j0.6)$$

$$\overline{I_R} = (262.43 - j\,196.8)\,Amps$$

Charging current at receiving end,

$$I_{C_1} = jV_R \frac{Y}{2} = j(63508.52)\left(\frac{4.5 \times 10^{-4}}{2}\right) = j1.289\,Amps$$

Line current, $\overline{I_L} = \overline{I_R} + \overline{I_{C_1}}$

$$I = 319.63\angle -34.82°\,Amps$$

Sending end voltage, $V_S = V_R + I_L Z$

$$= 63508.52\big((319.67\angle -34.82)(15 + j75)\big)$$

$$V_S = (8134.88 + j16944.5) \text{ volts}$$

$$V_S = 82885.38 \angle 11.59°$$

$$\theta_{V_S} = 11.79°$$

$$I_{C_2} = jV_S \frac{Y}{2}$$

$$\overline{I_{C_2}} = j(82885.38 \angle 11.79) \times \frac{4.5 \times 10^{-4}}{2}$$

$$\overline{I_{C_2}} = (-3.81 + j18.25) \text{ Amps}$$

Sending end current, $\overline{I_S} = \overline{I_L} + \overline{I_{C_2}}$

$$I_S = (262.43 - j182.53)(-3.81 + j18.25)$$

$$I_S = (258.62 - j164.27) \text{ Amps}$$

$$\theta_{I_S} = -32.42°$$

Sending end power factor, $\cos\phi_S = \cos(\theta_{V_S} - \theta_{I_S})$

$$\cos\phi_S = 0.716 \text{ lag}$$

Efficiency, $\%\eta = \dfrac{P_R}{P_R + 3I^2 R} \times 100$

$$= \frac{50 \times 10^6}{50 \times 10^6 + (3 \times 319.67^2 \times 15)} \times 100$$

$$\eta = 91.57\%$$

No load receiving end voltage, $V_{R_0} = V_S \left(\dfrac{\dfrac{-2j}{WC}}{R + jX_L - \dfrac{2j}{WC}} \right)$ (or) $\dfrac{V_S}{1 + \dfrac{1}{2}YZ}$

$$V_{R_0} = \frac{82882.38}{1 + \dfrac{1}{2} \times 4.5 \times 10^{-4} \angle 90° \times (15 + j15)}$$

$$V_{R_0} = 84304 \angle -0.196°$$

$$\% \text{ Regulation} = \frac{V_{R_0} - V_R}{V_R} \times 100$$

$$= \frac{84304 - 63508}{63508} \times 100$$

$$\% V_R = 32.74\%$$

16. **A 3-φ overhead transmission line delivers a load of 80MW at 0.8 P.f lagging and 220KV between the lines. Its total series impedance and shunt admittance /Ph are $200\angle 80°\Omega, 0.0013\angle 19°\mho$ respectively. Using nominal - T & nominal - Π method, calcuate (i) A, B, C, D constants of line (ii) sending end voltage (iii) sending end current (iv) sending end power factor (v) transmission efficiency (vi) regulation?**

Sol: *Given data:*

Series impedance, $Z = 200\angle 80°\Omega$

Shunt admittance, $Y = 0.0013\angle 90°\mho$

Power delivered, $P = 80MW$

Receiving end power factor, $\cos\phi_R = 0.8 \text{ lag} \Rightarrow \sin\phi_R = 0.6$

Receiving end voltage $= 200\,KV = V_R / Ph = 127017.05\,V$

Receiving end current, $I_R = \dfrac{80 \times 10^6}{\sqrt{3} \times 220 \times 10^3 \times 0.8} = 262.43\,Amps$

$$\overline{I_R} = I_R(\cos\phi_R - j\sin\phi_R) = 262.43(0.8 - j0.6)$$

$$= (209.94 - j157.45)\,Amps$$

$$= 262.42\angle -36.86°$$

For nominal - T method:

(i) $A = 1 + \dfrac{YZ}{2} = 1 + \dfrac{0.0013\angle 90° \times 200\angle 80°}{2} = 0.87 + 0.02j$

$\quad = 0.87\angle 1.48°$

$B = Z\left(1 + \dfrac{YZ}{4}\right) = 200\angle 80°\left(1 + \dfrac{0.0013\angle 90° \times 200\angle 80°}{4}\right)$

$$= 30.28 + j184.74$$

$$= 187.21 \angle 80.69° \, \Omega$$

$$C = Y = 0.0013 \angle 90° = 1.3 \times 10^{-3} j \text{ Siemen}$$

$$A = D = 0.87 \angle 1.48°$$

(ii) Sending end voltage, $V_S = AV_R + BI_R$

$$V_S = 127017.05(0.87 \angle 1.48°) + (30.28 + j\,184.74)(262.43 \angle -36.86°)$$

$$V_S = 150836.05 \angle 14.15° = 150.536 \angle 14.15° \text{ KV/Ph}$$

(iii) Sending end current, $I_S = CV_R + DI_R$

$$= (0.0013 \angle 90° \times 127017) + 0.87 \angle 1.48 \times (262.43 \angle 36.86°)$$

$$= 186.15 + j\,32.92$$

$$I_S = 189.04 \angle 10.03 \text{ Amps}$$

(iv) Sending end power factor, $\cos\phi_S = \cos(\theta_{V_S} - \theta_{I_S})$

$$= \cos(14.14 - 9.99)$$

(v) Transmission efficiency, $\eta = \dfrac{\text{Receiving end Power}}{\text{Sending end power}} \times 100$

$$= \dfrac{80 \times 10^6}{3 \times 150536.05 \times 189.04 \times 0.99} \times 100$$

$$\eta = 94.65 \, \%$$

(vi) $V_S = A\,V_{R_0}$ $\qquad\qquad [\because I_R = 0 \text{ at no load } V_R = V_{R_0}]$

$$V_{R_0} = \dfrac{150536.05 \angle 14.15°}{0.87 \angle 1.48}$$

% Voltage regulation $V_R = \dfrac{V_{R_0} - V_R}{V_R} \times 100 = \dfrac{173029.94 - 127017}{127017} \times 100$

For nominal – Π method:

(i) $A = 1 + \dfrac{YZ}{2} = 0.87 \angle 1.48°$

$A = D = 0.87 \angle 1.48°$, $B = Z = 200 \angle 80° = (34.72 + j\, 196.96)\,\Omega$

$C = Y\left(1 + \dfrac{YZ}{4}\right) = 1.46 \times 10^{-5} + j1.216 \times 10^{-3} = 1.216 \times 10^{-3} \angle 90.69°$ siemen

Sending end voltage $V_S = A V_R + B I_R$

$= 0.87 \angle 1.48 \times 127017 + 200 \angle 80° \times 262.43 \angle -36.86°$

$V_S = 153.72 \angle 14.59°$ KV / Ph

Sending end current, $I_S = C V_R + D I_R$

$= 1.216 \times 10^{-3} \angle 90.69° \times 127017 + 0.87 \angle 1.48 \times 262.43 \angle -36.86°$

$I_S = (185.62 \angle 6.88°)$ Amps

$\cos \phi_S = \cos(\theta_{V_S} - \theta_{I_S}) = \cos(14.59 - 6.88)$

$\cos \phi_s = 0.99$ lag

Transmission efficiency, $\eta = \dfrac{\text{Receiving end power}}{\text{Sendig end power}} \times 100$

$= \dfrac{80 \times 10^6}{3 \times 153.72 \times 10^3 \times 185.62 \times 0.99} \times 100$

$= 94.4\%$

$V_S = A V_{R_0}$

No load receiving end voltage, $V_{R_0} = \dfrac{153.72 \times 10^3}{0.87} = 176689.65\,V$

% Voltage regulation $= \dfrac{V_{R_0} - V_R}{V_R} \times 100 = \dfrac{176689.65 - 127017}{127017} \times 100$

$\%\eta = 39.1\%$

CHAPTER 3
Performance of Long Transmission Lines

3.1 Introduction

It is known that the line constants of a transmission line are uniformly distributed over the entire length of the line. So for it has been assumed that the line has lumped constants and line calculations made with such assumptions gave results. With reasonable accuracy. If such an assumption of lumped line constants is applied to long transmission lines, serious errors are introduced in the line performance calculations. Therefore, the performance calculations of long transmission line are made with line constants uniformly distributed ever the entire length of the line so that the results with fair degree of accuracy are obtained. Rigorous mathematical solution is required for the solution of such transmissionlines. The equivalent circuit of a 3ϕ long transmissionline is represented schematically as shown in below Fig.

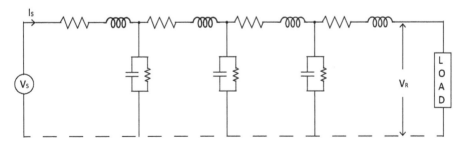

Fig. 3.1

3.2 Analysis of Long Transmission Lines By Rigorous Solution Method

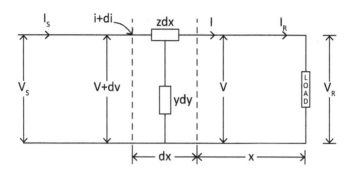

Fig. 3.2 Equivalent circuit of long transmission lines

For analysis, we shall take the receiving end as the reference for measuring distances. Take an elemental length 'dx' of the line a distance of 'x' from the receiving end say the voltage and current are $V+dV$ and $I+dI$ respectively.

Now, the series impedance of the element 'dx' of the line $= z\,dx$

Shunt admittance of element 'dx' of the line $= y\,dx$

Voltage across the elemental length in the direction of increasing 'x' is,

$$dV = I.Z\,dx$$

$$\frac{dV}{dx} = IZ \quad \rightarrow (1)$$

The current drawn by the element 'dx' is

$$dI = V.y\,dx$$

$$\frac{dI}{dx} = V.y \quad \rightarrow (2)$$

Differentiate equation (1) w.r.t to 'x' we get,

$$\frac{d^2V}{dx^2} = Z.\frac{dI}{dx}$$

$$= Z.V.y$$

$$\frac{d^2V}{dx^2} = Y.Z.V \quad \rightarrow (3)$$

The solution of the above differential equation is,
$$V = A_1 e^{\sqrt{YZ}\cdot x} + A_2 e^{-\sqrt{YZ}\cdot x} \qquad \rightarrow (4)$$

Differentiate equation (4) w.r.t to 'x' we get,
$$\frac{dV}{dx} = A_1 \sqrt{YZ}\cdot e^{\sqrt{YZ}\cdot x} - A_2 \sqrt{YZ}\cdot e^{-\sqrt{YZ}\cdot x}$$

$$\frac{dV}{dx} = \sqrt{YZ}\left\{A_1 e^{\sqrt{YZ}\cdot x} - A_2 \cdot e^{-\sqrt{YZ}\cdot x}\right\} \qquad \rightarrow (5)$$

From equation (1), we have,
$$I = \frac{1}{Z}\cdot\frac{dV}{dx}$$

$$= \frac{1}{Z}\cdot\sqrt{YZ}\left\{A_1 e^{\sqrt{YZ}\cdot x} - A_2 \cdot e^{-\sqrt{YZ}\cdot x}\right\} \qquad \rightarrow (6)$$

Equation (4) & (6) gives the expression for V and I in the form of unknown constants A_1 and A_2. The values of A_1 and A_2 can be obtained by applying receiving end conditions as under.

$$x = 0, \quad V = V_R \text{ and } I = I_R$$

Substituting these values in equation (4) & (6), we get
$$V_R = A_1 + A_2$$

and $\qquad \rightarrow (7)$

$$I_R = \sqrt{\frac{Y}{Z}}(A_1 - A_2) \qquad \rightarrow (8)$$

For a transmission line, $\sqrt{\frac{Z}{Y}}$ is a constant called "characteristic impedance (Z_c)" and $\sqrt{YZ}$ is another constant called "propagation constant (g)". Both are complex quantities.

From equation (7) & (8), we have,
$$A_1 = \frac{1}{2}\left(V_R + I_R \cdot \sqrt{\frac{Z}{Y}}\right)$$

$$A_1 = \frac{1}{2}(V_R + I_R \cdot Z_C)$$

$$A_2 = \frac{1}{2}\left(V_R - I_R \cdot \sqrt{\frac{Z}{Y}}\right)$$

$$A_2 = \frac{1}{2}(V_R - I_R \cdot Z_C)$$

Thus equation (4) & (6) becomes,

$$V = \left[\frac{1}{2}(V_R + I_R \cdot Z_C)\right]e^{\sqrt{YZ}\cdot x} + \left[\frac{1}{2}(V_R - I_R \cdot Z_C)\right]e^{-\sqrt{YZ}\cdot x}$$

$$V = \frac{1}{2}(V_R + I_R \cdot Z_C)e^{\gamma x} + \frac{1}{2}(V_R - I_R \cdot Z_C)e^{-\gamma x}$$

$$V = V_R\left(\frac{e^{\gamma x} + e^{-\gamma x}}{2}\right) + I_R Z_C\left(\frac{e^{\gamma x} - e^{-\gamma x}}{2}\right)$$

$$V = V_R \cosh \gamma x + I_R \cdot Z_C \sinh \gamma x \quad \rightarrow (9)$$

$$I = \frac{1}{Z_C}\left\{\left[\frac{1}{2}(V_R + I_R \cdot Z_C)\right]e^{\gamma x} - \left[\frac{1}{2}(V_R - I_R \cdot Z_C)\right]e^{-\gamma x}\right\}$$

$$= \frac{1}{Z_C}\left\{V_R\left(\frac{e^{\gamma x} - e^{-\gamma x}}{2}\right) + I_R Z_C\left(\frac{e^{\gamma x} + e^{-\gamma x}}{2}\right)\right\}$$

$$I = \frac{V_R}{Z_C}\sinh \gamma x + I_R \cosh \gamma x \quad \rightarrow (10)$$

Sending end voltage V_S and sending end current I_S can be obtained by substituting x l in the above equation,

$$V_S = V_R \cosh \gamma l + I_R Z_C \sinh \gamma l \quad \rightarrow (11)$$

$$I_S = I_R \cosh \gamma l + \frac{V_R}{Z_C}\sinh \gamma l$$

$$= \frac{V_R}{Z_C}\sinh \gamma l + I_R \cosh \gamma l \quad \rightarrow (12)$$

Now, $\gamma l = \sqrt{YZ} \cdot l$

$\quad = \sqrt{Yl \cdot Zl}$

$\quad = \sqrt{YZ}$

Where 'Z' is the total impedance of the line and 'Y' is the total admittance of the line.

Now, the expression for sending end voltage and sending end currents are,

$$V_S = V_R \cosh\sqrt{YZ} + I_R Z_C \sinh\sqrt{YZ} \text{ and}$$

$$I_S = \frac{V_R}{Z_C} \sinh\sqrt{YZ} + I_R \cosh\sqrt{YZ}$$

3.3 Calculation of A B C D Constants for Long Transmission Lines

Input voltage and current can be expressed in terms of o/p voltage and current as,

$$V_S = AV_R + BI_R \quad \rightarrow (1)$$

$$I_S = CV_R + DI_R \quad \rightarrow (2)$$

Input voltage and current of long Transmission line can be written as,

$$V_S = V_R \cosh\gamma l + I_R Z_C \cdot \sin\gamma l \quad \rightarrow (3)$$

$$I_S = V_R \cdot \frac{1}{Z_C} \sinh\gamma l + I_R \cosh\gamma l \quad \rightarrow (4)$$

Comparing equation (3) with (1)

$A = \cosh\gamma l$ $\quad\quad\quad\quad\quad\quad\quad\quad C = \dfrac{1}{Z_C}\sinh\gamma l$

$B = Z_C \sinh\gamma l$ $\quad\quad\quad\quad\quad\quad D = \cosh\gamma l$

Here, $A = D$ Symmetrical N/w $\quad\quad\quad AD - BC = 1$ passive N/w

3.4 Evaluation of A B C D Constants

A B C D constants can be evaluated by the following methods:

(1) **Convergent Series (Complex Angle) Method:**

In this method the hyperbolic *sin* and *cosine* are expressed in terms of their power series. The expressions are

$$\cosh x = 1 + \frac{x^2}{1!} + \frac{x^4}{4!} + \frac{x^6}{6!} + \ldots \text{ and}$$

$$\sinh x = x + \frac{x^3}{3!} + \frac{x^5}{5!} + \frac{x^7}{7!} + \ldots$$

$$= x\left(1 + \frac{x^2}{3!} + \frac{x^4}{5!} + \frac{x^6}{7!} + \ldots\right)$$

Now,

$$A = D = \cosh\sqrt{YZ}$$

$$A = D = 1 + \frac{YZ}{2} + \frac{Y^2Z^2}{24} + \frac{Y^3Z^3}{720} + \ldots$$

$$B = Z_C \sinh\sqrt{YZ} = Z_C\sqrt{YZ}\left\{1 + \frac{YZ}{6} + \frac{Y^2Z^2}{120} + \frac{Y^3Z^3}{5040} + \ldots\right\}$$

$$= \sqrt{\frac{Z}{Y}}\sqrt{YZ}\left\{1 + \frac{YZ}{6} + \frac{Y^2Z^2}{120} + \frac{Y^3Z^3}{5040} + \ldots\right\} \text{ and}$$

$$C = \frac{1}{Z_C}\sinh\sqrt{YZ} = \sqrt{\frac{Y}{Z}} \times \sqrt{YZ}\left\{1 + \frac{YZ}{6} + \frac{Y^2Z^2}{120} + \frac{Y^3Z^3}{5040} + \ldots\right\}$$

$$= Y\left\{1 + \frac{YZ}{6} + \frac{Y^2Z^2}{120} + \frac{Y^3Z^3}{5040} + \ldots\right\}$$

(2) **Convergent series Real Angle Method:**

In this method the hyperbolic *sin*&*cosine* are expanded by the trigonometrically formulas.

$$\gamma = \sqrt{YZ} = (\alpha + j\beta)$$

$$A = D = \cosh\sqrt{YZ}$$
$$= \cosh(\alpha + j\beta)$$
$$= \cosh\alpha \cosh j\beta + \sinh\alpha \sinh j\beta$$
$$= \cosh\alpha \cos\beta + j\sinh\alpha \sin\beta$$

$\alpha \rightarrow$ Attenuation constant
$\beta \rightarrow$ Phase constant

$$\sinh\sqrt{YZ} = \sinh\alpha \cosh jB + \cosh\alpha \sinh jB$$
$$= \sinh\alpha \cos\beta + j\cosh\alpha \sin\beta$$
$$B = Z_C\{\sinh\alpha \cos\beta + j\cosh\alpha \sin\beta\}$$
$$C = \frac{1}{Z_C}\{\sinh\alpha \cos\beta + j\cosh\alpha \sin\beta\}$$

3.5 Surge Impedance

Surge impedance of the line is defined as the square root of the ratio of series impedance (Z) and shunt admittance (Y) i.e.,

$$Z_0 = \sqrt{\frac{Z}{Y}}$$

$$Z = R + jX$$
$$Y = g + jWC$$

When the transmission system dealing with the surges or high frequencies, the losses are neglected means resistance and conductance i.e., zero.

$R = 0$ & $g = 0$ then
Character impedance becomes,

$$Z_C = \sqrt{\frac{Z}{Y}} = \sqrt{\frac{R + jW_L}{g + jW_C}}$$

$$Z_C = \sqrt{\frac{L}{C}} = Z_0$$

Surge impedance (Z_0) is the characteristic impedance of a loss less line:

The value of surge impedance for OHTL is 400Ω to 600Ω and that for the cable is around 40Ω to 60Ω. The lower value of surge impedance in case of cable is due to the relatively large capacitance and low inductance.

The surge impedance can also be evaluated by measuring the line impedance at the sending end when

1. The line is at receiving end is open circuited, i.e., $I_R = 0$
   Then $V_S = AV_R$, $I_S = CV_R$

   $$\therefore Z_{OC} = \frac{V_S}{I_S} = \frac{AV_R}{CV_R} = \frac{A}{C} \text{ and} \qquad \rightarrow (1)$$

2. The line at receiving end is short circuited, $V_R = 0$
   Then $V_S = BV_R$, $I_S = DV_R$

   $$\therefore Z_{SC} = \frac{V_{SC}}{I_{SC}} = \frac{BV_R}{DV_R} = \frac{B}{D} \text{ and} \qquad \rightarrow (2)$$

Multiply equation (1) & (2)

$$Z_{OC} \cdot Z_{SC} = \frac{A}{C} \times \frac{B}{D} [\because A = D]$$

$$Z_{OC} \cdot Z_{SC} = \frac{B}{C}. \qquad \rightarrow (3)$$

Substituting $B = Z_C \sinh\sqrt{YZ} = \sqrt{\frac{Z}{Y}} \sinh\sqrt{YZ}$

$$C = \frac{1}{Z_C} \sinh\sqrt{YZ} = \sqrt{\frac{Y}{Z}} \sinh\sqrt{YZ}$$

Equation (3) becomes,

$$Z_{OC} \cdot Z_{SC} = \frac{\sqrt{\frac{Z}{Y}} \sinh\sqrt{YZ}}{\sqrt{\frac{Y}{Z}} \sinh\sqrt{YZ}}$$

$$Z_{OC} \cdot Z_{SC} = \frac{Z}{Y}$$

Surge impedance $Z_0 = \sqrt{\frac{Z}{Y}}$

$$= \sqrt{Z_{OC} \cdot Z_{SC}}$$

$$\therefore Z_0 = \sqrt{Z_{OC} Z_{SC}}$$

3.6 Surge Impedance Loading (SIL)

Surge impendence loading is defined as the load that can be delivered by the line of negligible resistance where the load is at unity power factor.

Under this condition,

$$\text{Power delivered } P_R = \frac{V_R^2}{Z_0} MW$$

Where, $V_R \to$ Receiving end line voltage in 'KV'

$Z_0 \to$ Surge impedance in Ohms

$P_R \to$ Surge impedance loading

The power transmitted through a long transmission line can be increased in two ways.

1. Increasing the receiving end voltage V_R
2. Decreasing the surge impedance

Now-a-days the trend of employing higher and higher voltages for transmission. Therefore, this is the most commonly adopted method for increasing the power limit of heavily loaded long transmission lines. But these is a limit beyond which it is neighter economical nor practicable to increase the value of V_R.

Since the spacing between the conductors, which depends upon the line voltage employed can notreduced much. So the value of surge impedance (Z_0) can not be varied as such. However, some artificial means, such as series capacitors or shunt capacitors can be used to reduce the value of surge impedance (Z_0).

3.7 Wave Length & Velocity of Propagation of Wave

At any time the voltage and current vary harmonically along the line w.r.t to 'x'. The distance of the point consideration. A complete voltage or current cycle along the line corresponds to a change of 2π radians in the angular argument 'βx'. The corresponding line length defined as "wave length".

If 'β' expressed in rad/m,

Then the wave length $\lambda = \dfrac{2\pi}{\beta} m$

For a typical power transmission line,
$$g = 0 \ \& \ R \ll WL$$
$$\gamma = \sqrt{YZ} = (\alpha + j\beta)$$
$$= [jWC(R + jWL)]^{\frac{1}{2}}$$
$$= \left[(jWC)(jWL)\left(1 + \frac{R}{jWL}\right)\right]$$
$$= jW(LC)^{\frac{1}{2}}\left(1 - j\frac{R}{2WL}\right)^{\frac{1}{2}}$$
$$= jW(LC)^{\frac{1}{2}} + \frac{R}{2}\cdot\left(\frac{C}{L}\right)^{\frac{1}{2}}$$
$$= \frac{R}{2}\sqrt{\frac{C}{L}} + jW\sqrt{LC}$$
$$\gamma = \alpha + j\beta$$
$$\therefore \alpha = \frac{R}{2}\cdot\sqrt{\frac{C}{L}} \ \& \ \beta = W\sqrt{LC}$$

Time for phase change of 2π is $\frac{1}{f}$ cycles/sec

$t = \frac{1}{f}$ where $f = \frac{w}{2\pi}$ cycles/sec

During this time, the wave travels a distance equal to 'λ'. i.e., one wave length

Velocity of propagation of wave $V = \frac{\lambda}{1/f}$

$$V = \lambda f \ m/sec$$

For a loss less tr line,
$$R = 0 \ \& \ g = 0$$

Then $\gamma = \sqrt{YZ} = [(jWC)(jWL)]^{\frac{1}{2}}$
$$= jW\sqrt{LC}$$

$$\gamma = \alpha + j\beta$$

Such that $\alpha = 0$ & $\beta = W\sqrt{LC}$

$\therefore$ wave length $\lambda = \dfrac{2\pi}{\beta}$

$$= \dfrac{2\pi}{W\sqrt{LC}} \Rightarrow \dfrac{2\pi}{2\pi\sqrt{LC}}$$

$$\lambda = \dfrac{1}{f\sqrt{LC}} \; m \text{ and}$$

Velocity of propagation of wave $V = f\lambda$

$$V = f \cdot \dfrac{1}{f\sqrt{LC}}$$

$$V = \dfrac{1}{\sqrt{LC}} = \dfrac{1}{\left[2\times 10^{-7} \ln\dfrac{d}{r^1} \times \dfrac{2\pi\epsilon_0}{\ln\dfrac{d}{r}}\right]^{\frac{1}{2}}}$$

$$V = \dfrac{1}{\left[2\times 10^{-7} \times 2\pi\epsilon_0\right]^{\frac{1}{2}}} \qquad \left[\because \ln\dfrac{d}{r^1} \cong \ln\dfrac{d}{r}\right]$$

$$= 2.99 \times 10^8 \; m/sec$$

$$\cong 3 \times 10^8 \; m/sec = \text{velocity of light}$$

The actual velocity of propagation of wave (2.99×10^8) along the line would be somewhat less than velocity of light.

1. *Calculate the velocity of propagation for waves,*
   (a) In an overhead line of capacitance 0.147×10^{-10} F/m and inductance 0.75×10^{-6} H/m.
   (b) In a cable of inductance 0.75×10^{-6} H/m and capacitance 13.3×10^{-10} F/m.
   (c) Estimate the relative permittivity of insulating material in case (b).

Sol: *Given data:*

Capacitance of overhead line, $C_l = 0.147 \times 10^{-10}$ F/m

Inductance of overhead line, $L_l = 0.75 \times 10^{-6}$ H/m

Capacitance of cable $C_C = 13.3 \times 10^{-10}$ F/m

Inductance of cable $L_C = 0.75 \times 10^{-6}$ H/m

(a) velocity of propagation of wave in OHTL

$$v_L = \frac{1}{\sqrt{L_l \times C_L}}$$

$$= \frac{1}{\sqrt{0.75 \times 10^{-6} \times 0.147 \times 10^{-10}}}$$

$$= \frac{1}{\sqrt{1.1025 \times 10^{-17}}}$$

$$= 0.301 \times 10^9 \text{ m/sec}$$

(b) velocity of propagation of wave in cable

$$v_C = \frac{1}{\sqrt{L_C \times C_C}}$$

$$= \frac{1}{\sqrt{0.75 \times 10^{-6} \times 13.3 \times 10^{-10}}}$$

$$v_c = \frac{1}{\sqrt{9.975 \times 10^{-6}}}$$

$$= 31.662 \times 10^6 \text{ m/sec}$$

Relative permittivity
For a single phase two wire cable, we have

Inductance $L_C = 2 \times 10^{-7} \ln\left(\frac{d}{r}\right)$

$0.75 \times 10^{-6} = 2 \times 10^{-7} \ln\left(\frac{d}{r}\right)$

$$\ln\left(\frac{d}{r}\right) = \frac{0.75 \times 10^{-6}}{2 \times 10^{-7}}$$

$$= 3.75$$

Capacitance $C_C = \dfrac{2\pi\varepsilon_0}{\ln\left(\dfrac{d}{r}\right)}$

$$13.3 \times 10^{-10} = \frac{2\pi\varepsilon_0\varepsilon_r}{\ln\left(\dfrac{d}{r}\right)}$$

$$13.3 \times 10^{-10} = \frac{2\pi \times 8.854 \times 10^{-12} \times \varepsilon_r}{3.75}$$

$$\varepsilon_r = 89.775$$

2. **A 3φ single circuit transmission line is 400 km long. If the line rated for 400 KV and has line constants $R = 0.1\,\Omega/km$, $L = 1.26\,mH/km$, $C = 0.009\,\mu f/km$ and $G = 0$. Calculate the following:**
   (i) Surge impedance
   (ii) Velocity f surges
   (iii) If a surge of 500KV is incident at one end. Find the time to reach the open end.
   (iv) The voltage at the open end of the line.

Sol: Given data:

Length $l = 400\,km$

Rated voltage $V = 400\,KV$

Resistance $R = 0.1\,\Omega/km$

Inductance $L = 1.26\,mH/km$

Capacitance $C = 0.009\,\mu f/km$

Conductance $G = 0$

(i) surge impendence $Z_0 = \sqrt{\dfrac{L}{C}}$

$= \sqrt{\dfrac{1.26 \times 10^{-3}}{0.009 \times 10^{-6}}}$

$= 374.166 \, \Omega$

(ii) Velocity of surges neglecting R of line,

$$v = \dfrac{1}{\sqrt{LC}}$$

$= \dfrac{1}{\sqrt{1.26 \times 10^{-3} \times 0.009 \times 10^{-6}}}$

$= 296956.93 \, km/sec$

$= 2.97 \times 10^8 \, m/sec$

(iii) Time required for a surge of 500 KV to reach open end

$$t = \dfrac{length \, of \, the \, line}{velocity} = \dfrac{400}{2.97 \times 10^8}$$

$= 1.346 \, \mu sec$

(iv) Voltage at the open end $= 2 \times Applied \, voltage$

$= 2 \times 500$

$= 1000 \, KV$

3. A 3φ transmission line 200km long as the following constants resistance/Ph/km $= 0.16 \, \Omega$, reactance/Ph/Km $= 0.25 \, \Omega$, shunt admittance/Ph/km $= 1.5 \times 10^{-6}$ siemens. Calculate by regurous method the sending end voltage and current when the line is delivering a load of 20MW at 0.8 P.f lagging. The receiving end voltage is kept constant at 110KV?

Sol: *Given data:*

Length of the line $L = 200\,km$

Resistance/Ph/km $R = 0.16\,\Omega$

For 200 km, $= 0.16 \times 200 \times 10^3$

$\qquad = 32\,\Omega$

Reactance/Ph/Km $= 0.25\,\Omega$

For 200km, $X = 0.25 \times 200 \times 10^3$

$\qquad = 50\,\Omega$

Shunt admittance/Ph/km, $Y = 1.5 \times 10^{-6}$ siemens

For 200km, $Y = 1.5 \times 10^{-6} \times 200 \times 10^3$

$\qquad = 3.12 \times 10^{-4}\,s/km$

Series impedance, $Z = R + jX$

$\qquad = 32 \times j50$

$\qquad = 59.36 \angle 57.38°\,\Omega/km$

Receiving end voltage, $V_{R_{Ph}} = \dfrac{110 \times 10^3}{\sqrt{3}} = 63508.52\,V$

Receiving end current, $I_R = \dfrac{20 \times 10^6\,(Power)}{\sqrt{3} \times V_L \times \cos\phi} = 131.21\,Amps$

Taking V_R as reference phasor, we have,

$V_R = 63508.52 \angle 0°\,V$

$I_R = 131.21 \angle -36.87°\,V$

$YZ = (3.12 \times 10^{-4})(59.36 \angle 57.38°)$

$\qquad = 0.018 \angle 57.38°$

$YZ = 9.98 \times 10^3 + j0.015$

$Y^2Z^2 = (3.12 \times 10^{-4})^2 \times (59.36 \angle 57.38°)^2$

$\qquad = -1.43 \times 10^{-4} + j3.14$

Auxiliary line constants,

$$A = D = 1 + \frac{YZ}{2} + \frac{Y^2Z^2}{24}$$

$$= 1 + \frac{9.98 \times 10^3 + j0.015}{2} + \frac{\left(-1.43 \times 10^{-14} + j3.14\right)}{24}$$

$$= 1 + j0.13$$

$$= 0.014 \angle 7.77°$$

$$B = Y\left(1 + \frac{YZ}{6} + \frac{Y^2Z^2}{1120} + \ldots\right) \Omega$$

$$= 3.12 \times 10^{-4} \left[1 + \frac{0.018 \angle 57.38°}{6} + \left(\frac{-1.43 \times 10^{-4} + j3.14}{120}\right)\right]$$

$$= 3.12 \times 10^{-4} \angle 1.62°$$

$$V_S = AV_R + BI_R$$

$$= 1.014 \angle 7.79° (63508.52(0°)) + (59.48 \times 131.41 \angle -36.86°)$$

$$V_S = 65725.18 \angle 14.64° \text{ V}$$

Sending end current $I_S = CV_P + DI_R$

$$I_S = 3.126 \times 10^{-4} \angle 1.62 \ (63508.52 \angle 0°) + (1.014 \angle 70.78° \times 131.21 \angle 36.86)$$

$$I_S = 150.45 \angle -25.22° \text{ A}$$

CHAPTER 4
Power System Transients

4.1 Introduction

Transient phenomenon last in a power system for a very short period of time, ranging from a few μsec to 1 sec. Yet the study and understanding of this phenomenon is extremely important as during these transients, the system is subjected to the greatest stress from excessive over currents or voltages which depending upon their severity can cause extensive damage. In some extreme cases, there may be a complete shutdown of a plant or even a black-out of a whole area. Because of this, it is necessary to study the transients in power system network.

4.2 Types of System Transients

The main cause of momentary excessive voltages and currents are:

1. Lightning
2. Switching
3. Short circuit
4. Resonance conditions

Out of these, lighting and switching are the most common and usually most severe causes. Transients caused by short circuits or resonance conditions usually arise as secondary effects but may well lead to the plant breakdown in EHV systems.

Depending upon the speed of the transients, these can be classified as,

1. Surge phenomena (Extremely fast transients)
2. Short circuit phenomena (Medium fast transients)
3. Transient stability (Slow transients)

4.2.1 Surge Phenomena

This type of transient is caused by lighting (atmospheric discharges on overhead transmission lines) and switching physically, such a transient initiates an electromagnetic wave (surge) travelling with almost the speed of light ($3 \times 10^8 m/s$) on transmission lines. In a 150 km line, the travelling wave completes a round trip in 1ms. Thus, the transient phenomena associated with these travelling waves occur during the first few milliseconds after their initiation.

The reflection of surges at open line ends or at transformers which present high inductance leads to multiplicative effect on voltage build-up which may eventually damage the insulation of high voltage equipment with constant short circuit. The high inductance of the transformer plays the beneficial role of insulating the generator windings from transmission line surges. The travelling the generator windings from transmission line surges. The travelling charges in the surges are discharged to ground via lighting assisters without the initiation of a line short circuit these by protecting the equipment.

Selection of insulation level of various line equipment and transformers is directly related to the over voltages caused by surge phenomena.

4.2.2 Short-Circuit Phenomena

More than 50% short circuits take place on exposed overhead lines, owing to the insulation failure resulting from over voltages generated by surge phenomena, birds and other mechanical reasons. S.C.'s result from symmetrical (3ϕ) faults as well as unsymmetrical (LG, LL, LLG) faults. The occurrence of a symmetrical fault brings the power transfer across the line to zero immediately, where as the impact is only partial in case of unsymmetrical faults like surge phenomena, S.C.'s are also fully electric in nature. Their speed is determined by the time constants of the generator windings, which vary from a few cycles of 50Hz wave for the damper windings to around 4sec for the field winding. Therefore these transients will be sufficiently slower than the surge phenomena.

If S.C. currents allowed to persist, they may result in thermal damage to the equipment. Therefore, the faulty section should be isolated as quickly as possible. Most of the S.C's do not cause permanent damage. As soon as the fault is cleared. S.C. path is demonized and the insulation is restored.

4.2.3 Transient Stability

Whenever a S.C. takes place at any part of the integrated system, there is an instantaneous total or partial collapse of the bus voltages of the system. This also results in the reduction of the generator output. Since initially for some instants the i/p turbine power remains constant as there is always some time delay before the controllers can initiate corrective actions, each generator is subjected to positive accelerating torque. This condition if sustained for some time can result in the most severe type of transients namely the mechanical oscillations of the synchronous machines rotors. These electromechanical transients under extreme conditions lend to loss of synchronism for some or all of the machines, which implies that the power system has reached its transient stability limit. Once this happens, it may take, several hours for an electric system engineer to resynchronise such as 'blacked-out' system.

4.3 Attenuation & Distortion of Travelling Waves

Generally, in the analysis of travelling waves, attenuation and distortion are ignored as the lines are assumed to be lossless. But in practical applications, these occurs some losses, due to which it will become difficult to analyze the behavior of travelling waves. The resistance, leakage conductance of the line in addition to the corona causes losses in the transmission line. The losses due to line resistance dominates at low voltages whereas the losses due to corona dominates at high voltages.

Attenuation:

Attenuation is the reduction in magnitude of peak of travelling waves as it travels along the length of the line. Attenuation mainly occurs due to corona and the effect is more on positive waves than on negative waves.

Attenuation of a line can be found using foust and merger empirical formulae, which is given by

$$v = \frac{V}{1 + KxV}$$

Where,

- $v \to$ Surge voltage at distance 'x' from origin
- $V \to$ Surge voltage at a point of origin
- $x \to$ Distance travelled
- $K \to$ Attenuation constant
  - 0.00037 for chopped waves
  - 0.00019 for short waves
  - 0.0001 for long waves

Distortion:

Distortion is the change in the wave shape due to losses in the transmission line. As the wave travels along the length of the line, the losses cause wave the gets disturbed.

At low voltages, losses due to line resistance cause distortion, whereas at high voltages, corona causes distortion of the waves.

4.4 Reflection & Refraction of Travelling Waves

Whenever a travelling wave encounters change of impedance in the transmission line a part of the travelling wave reflects back to the source and a part of it gets refracted. This process takes place especially at the junction points, line – cable junctions and at terminations. Travelling wave's reflection and refraction phenomenon of travelling wave depends on the surge impedance of the transmission line. At junction points, line – cable junctions and at terminations, the surge impedance changes, which leads to the reflection and refraction of travelling waves.

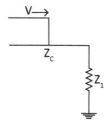

Consider a transmission line having characteristic impedance Z_C, terminated in an impedance Z_L as shown in figure.

Let a voltage 'V' be incidented on the transmission line. At the impedance Z_L, the wave encounters change in impedance and wave gets reflected back whereas a part of it gets refracted.

Let V_R and V_T be the magnitudes of reflected and refracted (transmitted) respectively. The magnitudes of these waves are given as,

For incident wave, $V = I Z_C$ $\rightarrow$ (1)

Where $I \rightarrow$ current wave associated with voltage wave (V)

For reflected wave, $V_R = I_R Z_C$ $\rightarrow$ (2)

Where $I_R \rightarrow$ current wave associated with voltage wave (V_R)

For refracted wave, $V_T = I_T Z_L$ $\rightarrow$ (3)

Where $I_T \rightarrow$ current wave associated with voltage wave (V_T)

Now, the algebraic sum of reflected and incident voltage is nothing but transmitted voltage.

The same thing holds good for current wave also. Therefore, we get,

$V_T = V + V_R$ $\rightarrow$ (4)

$I_T = I + I_R$ $\rightarrow$ (5)

Substituting the values of I, I_R, I_T and V_T from equation (1), (2), (3) & (4) in equation (5), we get

$$\frac{V_T}{Z_L} = \frac{V}{Z_L} - \frac{V_R}{Z_C}$$

$$\frac{V_T}{Z_L} = \frac{V}{Z_C} - \frac{(V_T - V)}{Z_C}$$

$$\frac{V_T}{Z_L} = \frac{2V}{Z_C} - \frac{V_T}{Z_C}$$

$$V_T \left(\frac{1}{Z_L} + \frac{1}{Z_C} \right) = \frac{2V}{Z_C}$$

$$V_T \left(\frac{Z_C + Z_L}{Z_L \cdot Z_C} \right) = \frac{2V}{Z_C}$$

$$V_T = \left(\frac{2Z_L}{Z_C + Z_L}\right) \times V \text{ volts} \qquad \rightarrow (6)$$

In the above equation, the term $\left(\dfrac{2Z_L}{Z_C + Z_L}\right)$ is the refraction coefficient.

Again from equation (4), we have

$$V_T = V + V_R$$

$$V_R = V_T - V$$

Substituting equation (6) in above equation,

$$V_R = \left(\frac{2Z_L}{Z_C + Z_L}\right) V - V$$

$$V_R = \left[\frac{2Z_L - (Z_C + Z_L)}{Z_C + Z_L}\right] V$$

$$V_R = \left(\frac{Z_L - Z_C}{Z_L + Z_C}\right) \times V \text{ volts}$$

In the above equation, the term $\left(\dfrac{Z_L - Z_C}{Z_L + Z_C}\right)$ is called as "Reflection coefficient".

∴ By using reflection & refraction coefficients, the reflected and refracted voltages can be found.

When the receiving end is open circuited, $I_R = 0$ & $Z_L = \infty$

Then,

The refraction coefficient $K_1 = \dfrac{2Z_L}{Z_L + Z_C}$

$$= \frac{2Z_L}{Z_L\left(1 + \dfrac{Z_C}{Z_L}\right)}$$

as $Z_L \rightarrow \infty$

$$= \frac{2}{\left(1 + \dfrac{Z_C}{\infty}\right)}$$

$$= \frac{2}{1+0}$$

$$K_1 = 2$$

The reflection coefficient $K_2 = \dfrac{Z_L - Z_C}{Z_L + Z_C}$

$$= \frac{Z_L\left(1 - \dfrac{Z_C}{Z_L}\right)}{Z_L\left(1 + \dfrac{Z_C}{Z_L}\right)}$$

as $Z_L \to \infty$

$$= \frac{Z_L\left(1 - \dfrac{Z_C}{\infty}\right)}{Z_L\left(1 + \dfrac{Z_C}{\infty}\right)}$$

$$= \frac{1-0}{1+0}$$

$$K_2 = 1$$

4.5 Termination of Transmission Line

4.5.1 *With Open End*

Consider a line with receiving end open circuited when switch S is closed a voltage and current waves of magnitudes V and I respectively travel towards the open-end. These waves are related by the equation.

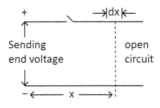

Fig. 4.1

$$\frac{V}{I} = Z$$

Where 'Z' is the characteristic impedance of the line. Consider the last element dx of the line. The electromagnetic energy stored by the element dx is given by $\frac{1}{2} L\,dx\,I^2$ and electrostatic energy in the element dx is $-c\,dx\,V$. Since the current at open end is zero, the electromagnetic energy vanishes and is transformed into electrostatic energy. As a result let the change in voltage be e; then

$$\frac{1}{2} L\,dx\,I^2 = \frac{1}{2} C\,dx\,e^2$$

$$\frac{e^2}{I^2} = \frac{L}{C}$$

$$e = IZ = V$$

This means the potential of open end is raised by V volts; then the total potential of the open end when the wave reaches this end is

$$V + V = 2V$$

The wave that starts travelling over the line when the switch S is closed could be considered as the incident wave and after the wave reaches the open end the rise in potential V could be considered due to wave which is reflected at the open end and actual voltage at the open end could be considered as the refracted wave or transmitted wave

Refracted wave = incident wave + Reflected wave

As soon as the incident wave I reaches the open end the current at the open end is zero. This means for an open end line the current wave is reflected with negative sign and coefficient of reflection unity. The variation of current and voltage waves over the line explained below.

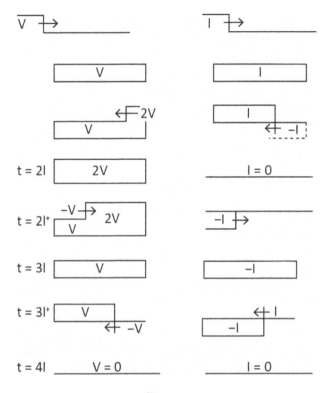

Fig. 4.2

After the voltage and current waves are reflected back from the open end they reach the source end, the voltage over the line becomes $2V$ and the current is zero. The voltage at source end cannot be more than source voltage V. therefore the voltage wave of $-V$ and current wave of $-I$ is reflected back into the line. After the waves have travelled through a distance of $4l$ where l is the length of the line. They would have wiped out both the current and voltage waves, leaving the line momentarily in its original state the above cycle repeats itself.

4.5.2 With Short Circuit End

Consider a transmission line short circuit at a distance x as shown in fig.

Now for an applied voltage V volts. A current $I = \dfrac{V}{Z_0}$ flows.

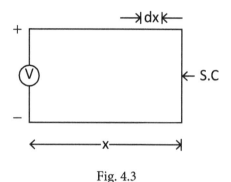

Fig. 4.3

As the transmission line possess capacitance and it is known that the voltage across the capacitor cannot be changed suddenly, the electrostatic energy stored by $C\,dx$ is converted into electromagnetic energy i.e.

$$\frac{1}{2}C\,dx\,V^2 = \frac{1}{2}L\,dx\,I_R^2$$

$$I_R = \frac{V}{Z_0} = I \left(\because \sqrt{\frac{L}{C}} = Z_0 \right)$$

Hence current at the point of short circuit becomes $I + I_R = I + I = 2I$. So a current of I amps flows towards the sending end in the reverse direction making the line voltage

$$V_L = V + (-I)Z_0 = V - IZ_0 = V - V = 0$$

But the voltage at sending end cannot be zero. Hence a voltage of V volts is again reflected into the line and the current now reaches to

$$I_L = 2I + \frac{V}{Z_0} = 2I + I = 3I$$

Till the current travels a distance of $3x$, again the energy conversion takes place, increasing the current to $4I$ and making line voltage to be '0'. Hence this process is repeated continuously and the line current is increased by I for each travel of x and finally. It reaches infinite value (if the line is a lossless one i.e. $R=0$). These variations in voltage and current are shown by the wave form.

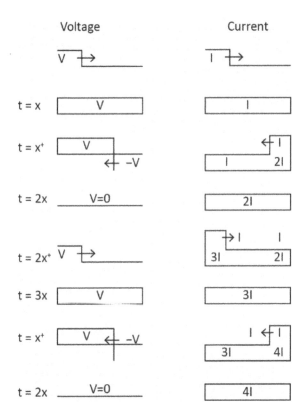

Fig. 4.4

4.5.3 With T-Junction

Let a line of natural impedance Z_C bifurcate into two branches of natural impedances Z_1 and Z_2. As far as the voltage wave is concerned, the transmitted portion will be same for both branches as they are in parallel but the refracted currents will be different as $Z_1 \neq Z_2$.

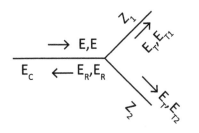

Fig. 4.5 Actual circuit

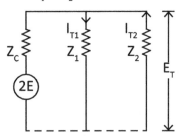

Fig. 4.6 Equivalent circuit

Let the incident wave be (E, I) travelling to the right, the reflected wave (E_R, I_R) travelling to the left and the transmitted waves (E_T, E_{T_1}) and (E_T, E_{T_2}) travelling towards right.

Let Z be an equivalent impedance of two parallel lines and equal to

$$Z = \frac{Z_1 Z_2}{Z_1 + Z_2}$$

Total impedance $Z_T = Z_C + Z$

$$= Z_C + \frac{Z_1 Z_2}{Z_1 + Z_2}$$

$$E_T = \frac{2E}{\frac{Z_1 Z_2}{Z_1 + Z_2} + Z_C} \times \frac{Z_1 Z_2}{Z_1 + Z_2} \Rightarrow E_T = 2E \times \frac{Z}{Z_C + Z}$$

$$= \frac{\frac{2E}{Z_C}}{\frac{1}{Z_C} + \frac{1}{Z_1} + \frac{1}{Z_2}}$$

$$I_{T_1} = \frac{E_T}{Z_1}$$

$$I_{T_2} = \frac{E_T}{Z_2}$$

Reflected voltage $E_R = E_T - E$

$$I_R = I_{T_1} + I_{T_2} - I$$

Problems

1. *A Rectangular surge of 1000KV incident on a overhead transmission line of surge impedance of 300Ω meets a junction of two cables of surge impedance 150Ω and 100Ω calculate the outgoing voltage and current on each cable and also the overhead line.*

A) Given data:

$$Z_c = 300\,\Omega$$

$$Z_1 = 150\,\Omega$$

$$Z_2 = 100\,\Omega$$

$$Z_C = Z_C + Z$$

$$= Z_C + \frac{Z_1 Z_2}{Z_1 + Z_2} = 300 + \frac{150 \times 100}{150 + 100} = 360\,\Omega$$

$$E_T = \frac{2E}{Z_T} \times \frac{Z_1 Z_2}{Z_1 + Z_2} = \frac{2 \times 1000}{360} \times \frac{150 \times 100}{150 + 100} = 333.2\,KV$$

$$I_{T_1} = \frac{E_T}{Z_1} = \frac{333.2 \times 10^3}{150} = 2220\,A$$

$$I_{T_2} = \frac{E_T}{Z_2} = \frac{333.2 \times 10^3}{100} = 3332\,A$$

$$E_R = E_T - E = 333.2 - 1000 = -666.8\,KV$$

Incident current $I = \dfrac{E}{Z_C} = \dfrac{1000 \times 10^3}{300} = 3333.3\,A$

Reflected current $I_R = I_T - I$

$$= I_{T_1} + I_{T_2} - I = 2220 + 3332 - 3333.3$$

$$= 2.2\,KA$$

2. A surge of 200 KV travelling on a line of surge impedance 400Ω reaches a junction of the line with two branch lines of surge impedance 600Ω and 400Ω respectively. Find surge voltage and current transmitted into each branch and line and also find reflected voltage and reflected current.

A) Given data:

$$Z_c = 400\,\Omega$$

$$E = 200\,KV$$

$Z_1 = 600 \Omega$

$Z_2 = 400 \Omega$

$Z_C = Z_C + Z$

$$= + \frac{Z_1 Z_2}{Z_1 + Z_2} = 400 + \frac{600 \times 400}{600 + 400} = 640$$

$$E_T = \frac{2E}{Z_T} \times \frac{Z_1 Z_2}{Z_1 + Z_2} = \frac{2 \times 200}{640} \times \frac{600 \times 400}{600 + 400} = 150 \, KV$$

$$I_{T_1} = \frac{E_T}{Z_1} = \frac{150 \times 10^3}{600} = 250 \, A$$

$$I_{T_2} = \frac{E_T}{Z_2} = \frac{150 \times 10^3}{400} = 37.5 \, A$$

$$E_R = E_T - E = 150 - 300 = -50 \, KV$$

Incident current $I = \dfrac{E}{Z_C} = \dfrac{200 \times 10^3}{400} = 500 \, A$

Reflected current $I_R = I_T - I$

$$= I_{T_1} + I_{T_2} - I = 250 + 375 - 500 = 125 \, A$$

4.6 Bewley's Lattice Diagrams

Consider a resistive load (R_L) connected to a generator (or) having resistance (R_g) through a transmission line of characteristic impedance (Z_C) as shown in fig (1).

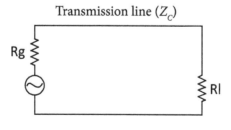

Fig.4.7.

If a voltage or current wave sent to the load by the generator then the wave reflects back to the generator after reaching the load (R_L). Again due to presence of resistance R_g at the generator side, the wave reflects back to the load. Hence the wave suffers from repeated reflections and to monitor these reflections, Bewley's lattice diagram is drawn which is also called zig-zag diagram.

In lattice diagram two axes are provided. a horizontal axis representing the distance along the system and a vertical axis showing the time. The passage of surges is represented by the lines whose slopes provide the time equal to the distance travelled. The reflected and transmitted waves can be achieved at any point of change impedance by multiplying the magnitude of incidence wave with their refraction and reflection coefficients. Lattice diagram of coefficient of current is negative of the reflection coefficient of voltage.

Consider a system as shown in figure (1), where a generator unit with internal resistance R_g is switched on line without attenuation having a characteristic or surge impedance Z_C, with load resistance R_l at its receiving end.

The reflection coefficient at the receiving end is given by

$$\alpha_R = \frac{R_L - Z_C}{R_L + Z_C}$$

The reflection coefficient at the sending end is given by

$$\alpha_S = \frac{R_g - Z_C}{R_g + Z_C}$$

Let 'T' be the time interval of surge from one end of the line to the other. Now as soon as the generator unit is switched on, a step voltage surge of infinite length travels down the line towards the receiving end. This is represented by a line (left to right) as shown in fig (2). When a surge reaches load end in time 'T' seconds, a surge of amplitude α_R is generated in the reflection process. This surge is then travelling towards the generator end and reaches the end in time '$2T$' seconds. This process continues endlessly.

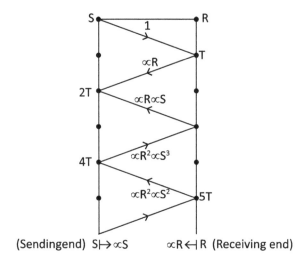

Fig.4.8.

After initiate reflections, the reflection voltage (V_R) becomes

$$V_R = (1+\alpha_R)\left[\frac{1}{1-\alpha_R \alpha_S}\right]$$

If $R_g = 0$

$$\alpha_S = \frac{R_g - Z_C}{R_g + Z_C} = \frac{-Z_C}{Z_C} = -1$$

∴ After infinite reflections, the reflection voltage is equal to incident voltage (which is unity) that means it reaches steady state after infinite reflections.

3. *A line of surge impedance of 400Ω is charged from a battery of constant voltage of 135 volts. The line is 300 meters long and is terminated in a resistance of 200 ohms. Plot reflection lattice and voltage across the terminating resistances.*

A) Surge impedance of transmission line $Z_C = 400\,\Omega$

Terminal resistance $R_L = 200\,\Omega$

Voltage of battery $E = 135\,V$

Reflection coefficient at the terminating end $\alpha_R = \dfrac{R_L - Z_C}{R_L + Z_C}$

$= \dfrac{200 - 400}{200 + 400}$

$= -\dfrac{1}{3}$

Reflection coefficient at the sending end $\alpha_S = \dfrac{R_g - Z_C}{R_g + Z_C} = \dfrac{-Z_C}{Z_C} = -1$

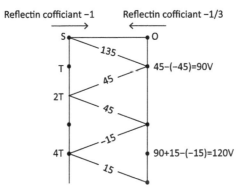

4. **A 400m long cable is short circuited at the remote end. A pulse source having resistance of 150Ω drives a 100V pulse. Having duration of $6\,\mu S$. If the characteristic resistance of the cable is 50Ω and the pulse velocity is $200\,m/\mu S$, sketch the voltage profile for first $8\,\mu S$ at the input of the line.**

A) Given data:

   Length of cable   400

   Source resistance $R_g = 150\,\Omega$

   Characteristic resistance of cable $R_C = 50\,\Omega$

   Pulse velocity $= 200\,m/\mu S = 200 \times 10^6\,m/sec$

   Pulse voltage $V = 100\,V$

   Pulse duration $= 6\,\mu S$

Time required for the wave to reach the remote end (S.C)

$$T = \frac{l}{V} = \frac{400}{2\times 10^8} = 2\,\mu S$$

Reflection coefficient at short circuit end $\alpha_R = \dfrac{R_L - R_C}{R_L + R_C}$

$$= -\frac{R_C}{R_C} = -1 \qquad\qquad \because [R_C = 0]$$

Reflection coefficient at sending end $= \dfrac{R_S - R_C}{R_S + R_C} = \dfrac{150 - 50}{150 + 50} = 0.5$

CHAPTER 5

Corona

5.1 Corona Phenomenon

Ionisation of air surrounding the power conductor is called "CORONA".

When voltage is applied across two conductors, whose spacing is large in comparison with their diameter, then the atmospheric air surrounding the conductor is subjected to electrostatic stresses. At low voltages (<100KV) there is no charge in the condition of atmospheric air around the conductor. However, when the potential difference is gradually increased, the field intensity around the conductor will increase. This will increase the velocity of electrons. Electrons travel with this velocity in space (or air) and collide with the molecules of air. As the velocity of the electrons increases, the collision process will increase and electrons are dislodge from the molecules of air. This collision process will keep on taking place in the air. After certain time, the electron avalanche wills takes place there by air gets ionized. This ionized air around the conductor will leads to corona. This corona can be identified by a faint luminous violet glow color and hissing noise.

If the conductors are perfectly uniform and smooth the glow will be uniform throughout the length of the conductor, otherwise rough points will appear brighter. If the spacing between the conductors is not much large as compared with their diameter, corona may bridge the conductors and cause flash over before violet glow observed. In order to avoid such flash over's between the conductors, the ratio of spacing between conductors to the radius of the conductors is much greater than 15.

$$\therefore \left(\frac{d}{r}\right) \gg 15$$

Corona is accompanied by a less of energy which increases very rapidly once the visual critical voltage is exceeded. Some energy will be lost due to corona and energy is dissipated in the form of light, heat, sound and chemical action. Due to corona, harmonics are produced. With this harmonics, non-sinusoidal voltage and current will flows through the lines.

Advantages and Disadvantages of Corona:

Corona is considered as something to be avoided because of energy loss associated with it and the distortion of waveform. But corona is considered beneficial because of the following advantages.

Advantages:

1. On the formation of corona, the sheath of air surrounding the conductor becomes conductive and there is virtual increase in conductor size i.e., diameter. Due to this virtual increase in conductor diameter, the maximum potential gradient or electrostatic stress is reduced. Thus probability of flash over is reduced and system performance will be improved.
2. Effects of transients produced by lightning and other causes are reduced. Since charges induced on the line by lightning or other causes will be partially dissipated as corona loss. In this way it acts as a safety value to the conductor (or Transmission line).

Disadvantages:

1. Energy required for the collision is taken from the supply. Hence some of the power will be lost for this process is called as "Corona power loss".
2. There is a non-sinusoidal voltage loop due to harmonic currents produced by corona and these may cause interfere with neighboring communication lines.
3. Owing to the formation of corona, ozone gas is produced which chemically react with the conductor and causes corrosion.

5.2 Critical Disruptive Voltage: (V_0)

It is the minimum phase to neutral voltage at which corona occurs is called "critical disruptive voltage". (Or)

Voltage at which the dielectric strength will be breakdown is called "critical disruptive voltage".

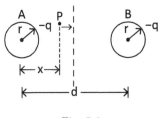

Fig. 5.1

Consider a 1-ϕ transmission line. Let 'r' be the radius of each conductor and 'd' be the distance betweenn the conductors, such that $d \gg r$.

Let '+q' be the charge at conductors A and '-q' be the charge at conductor 'B'. If the operating voltage is 'V', the potential of conductor 'A' w.r.t. to neutral plane 'N' will be 'V/2' and that of 'B' will be '-V/2'. Consider a point 'P' at a distance 'x' where electric field intensity need to be found. Bring unit positive charge at a point 'P'. The field due to 'A' will be repulsive and that of due to B will be attractive, these by the electric field intensity at 'P' due to both line charges will be additive and it will be

$$E_x = \frac{q}{2\pi\epsilon_0 x} + \frac{q}{2\pi\epsilon_0 (d-x)}$$

$$E_x = \frac{q}{2\pi\epsilon_0}\left(\frac{1}{x} + \frac{1}{d-x}\right)$$

The potential difference between the conductors is,

$$V = \int_{d-r}^{r} -E_x . dx$$

$$= -\int_{d-r}^{d} E_x \, dx$$

$$V = \int_r^{d-r} E_x \, dx$$

$$V = \frac{q}{2\pi\epsilon} \int_r^{d-r} \left(\frac{1}{x} + \frac{1}{d-x}\right) dx$$

$$= \frac{q}{\pi\epsilon_0} \left[(\ln x)_r^{d-r} + (\ln(d-x)(-1))_r^{d-r} \right]$$

$$= \frac{q}{2\pi\epsilon_0} \left[(\ln(d-r) - \ln r) - (\ln(d-d+r) - \ln(d-r)) \right]$$

$$= \frac{q}{2\pi\epsilon_0} \left[\ln(d-r) - \ln r - \ln r + \ln(d-r) \right]$$

$$= \frac{q}{2\pi\epsilon_0} [2\ln(d-r) - \ln r]$$

$$= \frac{q}{\pi\epsilon_0} \times \ln\left(\frac{d-r}{r}\right)$$

Since 'r' is very small as compared to 'd', so $(d-r) \cong d$

$$\therefore V = \frac{q}{\pi\epsilon_0} \ln\left(\frac{d}{r}\right)$$

$$q = \frac{\pi\epsilon_0}{\ln\left(\frac{d}{r}\right)} \cdot V \quad\quad\quad \rightarrow (1)$$

Now, the gradient at point 'P' at a distance 'x' from the centre of the conductor 'A' is given by

$$g_x = E_x = \frac{q}{2\pi\epsilon_0} \left[\frac{1}{x} + \frac{1}{d-x}\right]$$

$$g_x = \frac{q}{2\pi\epsilon_0} \cdot \frac{d}{x(d-x)} \rightarrow (2)$$

Substitute equation (1) in equation (2)

$$g_x = \frac{1}{2\pi\epsilon_0} \cdot \frac{\pi\epsilon_0}{\ln\left(\frac{d}{r}\right)} \cdot V \cdot \frac{d}{x(d-x)}$$

$$g_x = \frac{V \times d}{2x(d-x)\ln\frac{d}{r}}$$

Now $g_x = \dfrac{V_0 d}{x(d-x)\ln\dfrac{d}{r}}$ where $V_0 = \dfrac{V}{2}$

Where V_0 is the line to neutral voltage (i.e. phase voltage).

From the above expression it is clear that for a given transmission system the gradient increases as 'x' decreases. This gradient is maximum when $x = r$, i.e. the surface of the conductor and which is given as 'g_{max}'.

$$g_{max} = \frac{V_0 \times d}{r(d-r)\ln\dfrac{d}{r}}$$

As Insert Formula $d \gg r, (d-r) \simeq d$

$$g_{max} = \frac{V_0 \times d}{r.d.\ln\left(\dfrac{d}{r}\right)}$$

$$g_{max} = \frac{V_0}{r\ln\left(\dfrac{d}{r}\right)}$$

$$V_0 = g_{max}.r.\ln\left(\frac{d}{r}\right)$$

This voltage corresponds to the gradient at the surface equal to breakdown strength of air. Hence the voltage is called "Disruptive voltage".

The maximum gradient at the surface of the conductor corresponds to dielectric strength which is equal to 30 KV/cm peak at $25°C$ & 76 cm Hg. rms value of dielectric strength $= \dfrac{30}{\sqrt{2}}$ KV/cm $= 21.21$KV/cm. It is denoted as g_0.

$$V_0 = g_0\, r \ln\left(\frac{d}{r}\right)$$

At any other temperature and pressure

$$V_0 = \delta\, g_0\, r\ln\left(\frac{d}{r}\right)$$

$$\delta = \frac{3.92b}{273+t}$$

Where $\delta \to$ Air density correction factor
$b \to$ Barometric pressure
$t \to$ Temperature

In deriving the above expression, the assumption is made that the conductor is solid and the surface is smooth. For higher voltages ACSR conductors are used. The potential gradient for such conductor will be greater than for the equivalent smooth conductors. So that breakdown voltage for a stranded conductor will be less than the smooth conductor. The irregularities on the surfaces of such conductor are increased further because of the deposition of dust and dirt on its surface and breakdown voltage is further reduced. So, considering these irregularities into account, the critical disruption voltage becomes,

$$V_0 = m_0\, \delta\, g_0\, r \ln\left(\frac{d}{r}\right) \text{Kv/ph}$$

Where 'm_0' is inequality factor and it varies depending upon surfaces of the conductors.

5.3 Critical Visual Voltage (V_v)

It is the minimum phase to neutral voltage at which corona appears is called "critical visual voltage".

The voltage at which the dielectric strength of air gets breakdown, the corona phenomenon starts but it is not visible because the charged ions in the air must receive some finite energy because further ionization by collisions.

Peek's states that critical disruptive voltage must be exceeded that the stress is greater than breakdown value of air up to a distance of $0.3\sqrt{r\delta}$ from the centre of the conductor. Thus visual corona occurs when the breakdown is attained at a distance $(r+0.3\sqrt{r\delta})$ cm from the axis. This requires that the voltage to neutral be $\left(1+\frac{0.3}{\sqrt{\delta r}}\right)$ times the critical disruptive voltage.

According to peek's, the distance between g_v and g_0 is knows as "Energy distance". This distance equal to $(r+0.301\sqrt{r})$ for 2 parallel conductors and $(r+0.308\sqrt{r})$ for coaxial conductors.

From this it is clear that g_v is not constant as g_0 and is a function of the size of conductor.

$$g_v = g_0 \delta \left(1+\frac{0.3}{\sqrt{r\delta}}\right) \text{ KV/cm for 2 parallel conductors}$$

If V_V is critical visual voltage, then

$$V_V = g_v . r . \ln\left(\frac{d}{r}\right)$$

$$V_V = g_0 \delta \left(1+\frac{0.3}{\sqrt{r\delta}}\right) r \ln\left(\frac{d}{r}\right)$$

If irregularity factor is taken into account

$$V_V = m_V \, g_0 \, \delta \left(1+\frac{0.3}{\sqrt{r\delta}}\right) r \ln\left(\frac{d}{r}\right) \text{ KV}$$

Since the surface of the conductor is irregular, the corona does not start simultaneously on the whole surface but it takes place at different points of the conductor which are pointed and this is known as "local corona". For this $m_V = 0.72$ and for decided / general corona $m_V = 0.82$.

5.4 Corona Loss: (P)

The ions produced by the electric field result in space of charges which move round the conductor. The energy required for the charges to remain in motion is derived from the supply system. The space surrounding the conductor is loss. In order to maintain the flow of energy over the conductor in the field where in this additional energy would have been otherwise absent. It is necessary to supply this additional loss from the supply system. This additional power is referred to as "corona loss".

Peek's made a number of experiments to study the effect of various parameters on the corona loss and he deduced an empirical relation as

$$P = 241 \times \left(\frac{f+25}{\delta}\right)\sqrt{\frac{r}{d}}(V_p - V_0)^2 \times 10^{-5} \text{ Kw/km/ph}$$

This equation is for fair weather conditions.

For foul (or) stormy weather conditions. The corona loss;

$$P = 241 \times \left(\frac{f+25}{\delta}\right)\sqrt{\frac{r}{d}}(V_p - 0.8V_0)^2 \times 10^{-5} \text{ Kw/km/ph}$$

Where f → Supply frequency, δ → Air density correction factor

r → Radius of conductor, d → Spacing between the conductors

V_p → Operating voltage (Phage voltage)

V_0 → Critical Disruptive voltage

Limitation of Peek's formula:

1. If the supply frequency lays between 25 Hz to 120 Hz, then only it gives exact results.
2. Conductor radius is greater than 0.25 cm
3. The ratio $\frac{V_p}{V_0} > 1.8$ then it will give good results.

When the ratio $\frac{V_p}{V_0} < 1.8$, thus formula will helpless. Even under this limitation, Petoson's given a formula which holds good for ratio loss than 1.8 which is

$$P = \frac{21 \times 10^{-6} \times f V_p^2}{\left(\ln\frac{d}{r}\right)^2} \times F \text{ Kw/km/Ph}$$

Where F corona loss function and it varies as given below.

$\frac{V_p}{F}$	0.6	0.8	1.0	1.2	1.4	1.6	1.8	2.0	2.2
F	0.012	0.018	0.05	0.08	0.3	1.0	3.0	6.0	8.0

5.5 Factors Affecting Corona Loss

The following are the factors that affect corona loss on overhead transmission lines;

1. Electric Factors
2. Atmospheric Factors
3. Factors connected with the conductors

Corona loss can be expressed as,

$$P = 241 \times \left(\frac{f+25}{\delta}\right) \sqrt{\frac{r}{d}} (V_p - v_0)^2 \times 10^{-5} \text{ Kw/km/Ph}$$

1. Electrical Factors:

(a) Supply Frequency:

From the corona loss equation, frequency is proportional to the corona loss. As the frequency increases, the corona loss will also increase.

(b) Field around the conductor:

We know that $E = \dfrac{V}{r \ln\left(\dfrac{d}{r}\right)}$

From the above equation, as the operating voltage is increased, field around the conductor is also increases. When field is increases, velocity of electrons increases, thereby collision process will increase, hence corona losses will increase. As V decreases, E also decreases there by corona loss will also decrease.

(c) Shape of the waveforms:

When the corona occurs, harmonic currents will flow through the conductors. These harmonic currents will disturb the shape of waveform.

2. Atmospheric Factors:
(a) Pressure, Temperature:

As we know that the air density correction factor δ is

$$\delta = \frac{3.92b}{273+t}$$

Pressure and temperature will affect the corona loss. If temperature increases, air density correction factor decreases thereby corona loss will increase. If barometric pressure increases, air density correction factor increases thereby corona loss will decrease.

(b) Dust, Drain, Snow, Hail Effect:

The particles of dust clog to the conductors; thereby the critical voltage for local corona reduces which increase corona loss. Similarly, the bad atmospheric conditions such as rains, snow and hail storm reduces the critical disruptive voltage and hence increase the corona loss.

3. Factors connected with conductors:
(a) Diameter of conductors:

From the expression of corona losses, it can be seen that the conductor size appears at two places,

$$\text{Loss} \propto \sqrt{\frac{r}{d}} \text{ and}$$

$$\text{Loss} \propto (V_p - V_0)^2$$

From the first relation, loss is proportional to the size of the conductor. Larger the diameter of conductor, larger will be the loss. But from second relation, V_0 is approximately directly proportional to the size of the conductor. Hence larger the size of the conductor, larger will be the critical disruptive voltage and hence smaller will be the factor $(V_p - V_0)^2$. So corona loss will decrease. The effect of second relation is more than the first relation on corona loss. Hence larger the size of the conductor, lower will be the corona loss.

(b) **No. Of conductors per phase:**

If the no. of conductors per phase (i.e., Bundled conductor) Increases, GMR will increase,

$$E = \frac{V}{GMR \ln\left(\frac{GMR}{GMD}\right)}$$

From the above equation, if GMR increases, electric field intensity will decrease. If field intensity decreases, the collision process will decease there by corona loss will decrease.

(c) **Shape of conductor:**

Cylindrical conductors have uniformity in the distribution of the field. Hence corona losses decrease.

5.6 Methods of Reducing Corona

The corona can be reduced by the following methods.

(I) **By increasing conductor size:**

When the size of the conductor is increased, the voltage at which corona occurs and appears also increases. As the size of the conductor is increased, the electric field intensity reduces which in turn reduces the effect of corona. However, increase in size of conductor will increase the cost of conductors, mechanical stresses on the insulator etc. Thus, size of the conductor can't increase to a larger value in order to avoid early occurrence of corona, Thereby reducing the effect of corona.

(ii) **By increasing spacing between the conductors:**

By increasing the spacing between the line conductors, the electrostatic stress between the two conductors decreases and hence the air between the conductors gets ionized at a higher voltage i.e., the value of V_0 and V_V increases and effect of corona is reduced. However, the spacing between the conductors cannot be increased to a large value as this effects the size of the tower, weight of the supporting structures and the land occupied by the tower. Hence, an optimum value of sparing should be chosen so that V_0 increases and the effect of corona are reduced.

(iii) By using Bundled conductors:

For the same power rating by using bundled conductors the diameter of the conductor increases virtually. By increasing the diameter of the conductor, Electric field intensity will decrease which in terms decreases the effect of corona.

Problems on Corona:

1. A 3-φ line has conductors 2 cm in diameter spaced equilateral 1m apart. If the dielectric strength of air is 30 KV/cm (Max) find the critical disruptive voltage for the line. Take air density factor $\delta = 0.952$ and irregularity factor $m_0 = 0.9$?

Ans: Given data:

Diameter of the conductor, $D = 2\,cm$

Dielectric strength of air is, $g_0 = 30$ kv/cm (Maximum)

Dielectric strength of air is RMS value, $g_0 = 21.2$ kv/cm

Air density factor, $\delta = 0.952$

Irregularity factor, $m_0 = 0.9$

Radius of the conductor, $r = \dfrac{D}{2} = \dfrac{2cm}{2} = 1\,cm$

Spacing between the conductor, $d = 1\,m$
$$= 100\,cm$$

Critical disruptive voltage, $V_0 = m_0 g_0 \delta r \ln\left(\dfrac{d}{r}\right)$

$$V_0 = 0.9 \times 21.2 \times 1 \times \ln\left(\dfrac{100}{1}\right) \times 0.952$$

$$V_0 = 83.64\ \text{KV/Ph}$$

So line, critical disruptive voltage, $V_0 = \sqrt{3} \times 83.64$
$$V_0 = 144.88 \text{ KV}$$

2. Find the critical disruptive voltage and critical visual voltage for local and general corona on 3-ϕ over head transmission line and consisting of 3 standard copper conductors and 2.5 m a part at the corners of an equilateral triangle. Air temperature and pressure are 21 and 73.5 cm of Hg respectively. Conductor diameter is 1.8 cm. irregularity factor 0.85 and surface factors for 10 cal& general corona 0.7 & 0.8 respectively. Breakdown strength of air is 21.2 KV/cm?

Ans: Given data:

Spacing between the conductors, $d = 2.5$ m
$$= 250 \text{ cm}$$

Diameter of the conductor, $D = 1.8$ cm

Then, Radius, $r = \dfrac{D}{2} = \dfrac{1.8}{2} = 0.9$ cm

Temperature, $t = 21°C$
Pressure = 73.5 cm of Hg
Breakdown strength of air, $g_0 = 21.2$ KV/cm
Irregularity factor, $m_0 = 0.85$
Surface factor for local corona, $M_V = 0.7$
and for general corona, $m_V = 0.8$

Air density factor, $f = \dfrac{3.92b}{273+t} = \dfrac{3.92 \times 73.5}{273+21} = 0.98$

Critical disruptive voltage, $V_0 = m_0 \delta g_0 r \ln\left(\dfrac{d}{r}\right)$

$V_0 = 0.85 \times 0.98 \times 21.2 \times 0.9 \times \ln\left(\dfrac{250}{0.9}\right)$

$= 89.43$ KV/Ph

(i) For local corona:

$$V_V = m_V g_0 \delta \left(1 + \dfrac{0.3}{\sqrt{rs}}\right) r \ln\left(\dfrac{d}{r}\right)$$

$$= 0.7 \times 21.2 \times 0.98 \left(1 + \frac{0.3}{\sqrt{0.9 \times 0.98}}\right) 0.9 \ln\left(\frac{250}{0.9}\right)$$

$V_v = 97.125$ KV/Ph

(ii) for general corona:

$$V_v = m_v g_0 \delta \left(1 + \frac{0.3}{\sqrt{\delta r}}\right) \times r \times \ln\left(\frac{d}{r}\right)$$

$$= 0.8 \times 21.2 \times 0.98 \left[1 + \frac{0.3}{\sqrt{0.98 \times 0.9}}\right] \times \ln\left[\frac{250}{0.9}\right]$$

$V_v = 111.05$ KV/Ph

3. **A 132 KV line with 1.956 cm die conductor is built so that corona takes place if the line voltage exceeds 210 KV. If the value of potential gradient at which ionization occurs can be taken as 30 KV/cm. Find the spacing between the conductors?**

Ans: Given data:

Disruptive voltage, $V_0 = 210$ KV

Diameter of the conductor, $D = 1.956$ cm

then, radius, $r = \dfrac{D}{2} = 0.978$ cm

Phase disruptive voltage, (V_0)Ph $= \dfrac{210 \times 10^3}{\sqrt{3}} = 12124.55$ V

Dielectric strength (or) potential gradient, $g_0 = 30$ KV/cm (Max)

g_0 (RMS) $= 21.2$ KV/cm

Assume, surface irregularity factor & air density factor M_0 as one, i.e. M_0

$$V_0 = M_0 g_0 \delta r \ln\left(\frac{d}{r}\right)$$

$$121243.55 = 1 \times 1 \times 21.2 \times 0.978 \times \ln\left(\frac{d}{0.978}\right)$$

$$\ln\left(\frac{d}{0.978}\right) = 5.847$$

$$\frac{d}{0.978} = e^{5.847}$$

$$d = 3.38 \text{ M}$$

4. **A 110 KV, 3-φ 50 Hz, transmission line 175 km long consists of 3 standard copper conductors of 1 cm diameter speed in 3 meters lines are arranged in delta arrangement. Temperature and pressure are taken at $26°C$ and 74 cm of Hg respectively. Assume surface irregularity factor 0.85 and irregularity factors for local and general corona 0.72 & 0.82 find (i) disruptive voltage (ii) visual voltage for local & general corona (iii) corona loss for fair & wet weather conditions?**

Ans: Given data:

Line voltage $V_L = 110$ KV, Spacing $d = 3m = 300$ cm

Length of transmission line, $l = 175$ km

Diameter of conductor, $D = 1$ cm, $r = \dfrac{D}{2} = 0.5$ cm

The conductors are arranged in delta connection.

So line voltage = phase voltage

Temperature, $t = 26°C$

Pressure, $b = 74$ cm of Hg

Surface irregularity factor, $M_0 = 0.85$

Irregularity factors for local corona, $M_V = 0.72$

for general corona, $m_V = 0.82$

$$\delta = \frac{3.92 \times b}{273 + t} = \frac{3.92 \times 74}{273 + 26} = 0.97 \rightarrow \text{air density factor}$$

(i) Disruptive voltage, $V_0 = m_0 \delta g_0 r \ln\left(\dfrac{d}{r}\right)$

$$= 0.85 \times 0.97 \times 21.2 \times 0.5 \times \ln\left(\frac{300}{0.5}\right)$$

$$V_0 = 55.90 \text{ KV/Ph}$$

(ii) Critical visual voltage, $V_V = ?$

(a) for general corona:

$$V_V = m_v g_0 \delta \left(1 + \frac{0.3}{\sqrt{r\delta}}\right) r \ln\left(\frac{d}{r}\right)$$

(b) For local corona :

$$V_V = M_v g_0 \delta \left(1 + \frac{0.3}{\sqrt{r\delta}}\right) r \ln\left(\frac{d}{r}\right)$$

$$= 0.72 \times 21.2 \times 0.97 \left(1 + \frac{0.3}{\sqrt{0.5 \times 0.97}}\right) 0.5 \ln\left(\frac{300}{0.5}\right)$$

$V_V = 67.75$ KV/Ph

(iii) Corona loss,

$$P = 241 \times \left(\frac{f+25}{ä}\right) \times \sqrt{\frac{r}{d}} \times (V_P - V_0)^2 \times 10^{-5} \text{ KW/km/Ph}$$

For 125 km,

$$P = 241 \times \left(\frac{50+25}{0.97}\right) \times \sqrt{\frac{0.5}{300}} \times (110 - 55.90)^2 \times 10^{-5} \times 175$$

$P = 3896.39$ KW

This loss is for fair weather conditions
For wet weather conditions,
Corona loss,

$$P = 241 \times \left(\frac{f+25}{f}\right) \times \sqrt{\frac{r}{d}} \times (V_P - 0.8V_0)^2 \times 10^{-5} \text{ KW/km/Ph}$$

$$P = 241 \times \left(\frac{50+25}{0.97}\right) \times \sqrt{\frac{0.5}{300}} \times (110 - 0.8 \times 55.9)^2 \times 10^{-5} \times 175$$

$P = 5678.21$ KW/Ph

5. **A 3-φ, 50 Hz, 132 KV transmission line consists of conductors 1.17 cm diameter & the space equilaterally at a distance of 3 meters,**

the line conductors have smooth surface with value of 0.96. The barometric pressure is 72 cm of Hg & temperature is $20°C$. Find corona loss for fair & foul weather conditions?

Ans: Given data:

Diameter of the conductor, $D = 1.17$

then radius, $r = \dfrac{D}{2} = \dfrac{1.17}{2} = 0.585$ cm

Spacing between the conductor, $d = 3$ m $= 300$ cm
Surface irregularity factor, $M_0 = 0.96$
Temperature, $t = 20°C$
Pressure, $P = 72$ cm of Hg, $g_0 = 21.2$ KV/cm

Disruptive voltage, $V_0 = M_0 g_0 \delta r \ln\left(\dfrac{d}{r}\right)$

Air density factor, $\delta = \dfrac{3.92 \times b}{273 + t} = \dfrac{3.92 \times 72}{273 + 20} = 0.96$

$V_0 = 0.96 \times 21.2 \times 0.96 \times 0.585 \ln\left(\dfrac{300}{0.585}\right)$

$= 71.32$ KV/Ph

Corona loss for fair weather conditions,

$P = 241 \times \left(\dfrac{f+25}{\delta}\right) \times \sqrt{\dfrac{r}{d}} \times (V_p - 0.8 V_0)^2 \times 10^{-5}$

$P = 0.198$ KW/Ph/Km

For foul weather conditions,

Corona loss, $P = 241 \times \left(\dfrac{50+25}{0.96}\right) \times \sqrt{\dfrac{0.585}{300}} \times (76.21 - 0.8 \times 71.32)^2 \times 10^{-5}$

$= 3.05$ KV/Ph/km

CHAPTER 6
Mechanical Design of Overhead Transmission Lines

6.1 Overhead Line Insulators

The overhead line conductors are not covered with any insulating covering / coating. The line conductors are therefore secured to the supporting structures by means of insulating fixtures called "Insulators". In order that there is no current leakage to the earth through the supports. Insulators are mounted on the cross-arms and the line conductors are attached to the insulators so as to provide the conductors proper insulation and also provide necessary clearances between conductors and metal work. Insulators must provide proper insulation and necessary clearances against the highest voltage in worst atmospheric conditions to which the line is subjected.

Importances of an Insulator:

- It can prevent the leakage currents to the ground through steel tower and cross-arm.
- It can prevent short circuit between two phase conductors
- It can provide mechanical strength to conductors so that it can stretch more between towers.

6.2 Properties of Insulators

Materials used for an insulators should posses the following properties.

1. **High Mechanical Strength:** It should have high mechanical strength so as to bear the load due to weight of conductor, wind force and formation of ice if any.

2. **High insulation resistance:** It should have high insulation resistance in order to prevent leakage currents to earth.
3. **High Relative Permittivity:** It should have high relative permittivity so as to provide high dielectric strength.
4. **Non-Hygroscopic:** It should be non-hygroscopic means it should not absorb moisture (dust & dirt) easily from air.
5. **Pours & Impervious:** It should not be Pours such that it should not allow the dust to enter to enter into the body of the insulator. And it should not be impervious to the fluids and gases in the atmosphere.
6. **Rupture strength to Flash over voltage:** It should have high ratio of rupture strength to flash over voltage.
7. **Temperature variations:** It should have ability to withstand large temperature variations. i.e., it should not crack when subjected to high temperatures during summer and low temperatures during winter. The dielectric strength should remain unaffected under different conditions of temperature and pressure.

6.3 Materials Used for Insulators Are Two Types

1. Porcelain Insulators
2. Glass Insulators

Porcelain	Glass
1. It is a mixture of materials. i.e., clay (50%) + Feldspar (30%) + Silica (20%)	1. It is a single material which is Silica
2. It is not Homogeneous material.	2. It is Homogeneous material.
3. It is Non-hygroscopic material.	3. It is also a non-hygroscopic material but after few days of rapid use, it will become hygroscopic.
4. It can be used for 400 to 765 KV.	4. It can be used for up to 33 KV.
5. It is used for Transmission.	5. It is used for distribution.
6. Dielectric strength is 60KV/cm	6. Dielectric strength is 120 KV/cm

Porcelain	Glass
7. Fault can't be easily detected as it is not transparent.	7. Fault can be easily detected as it is transparent.
8. It has high mechanical strength.	8. It has low mechanical strength.

Toughened glass is also sometimes used for insulator because it has higher dielectric strength which makes it possible to make use of single piece construction, whatever is the operating voltage. It has lower coefficient of thermal expansion and strains due to temperature changes are minimized. The major drawback of glass is that moisture condenses very easily on its surface and hence its use is limited to 33 KV.

The design of the insulators is such that the stress due to contraction and expansion in any part of the insulator does not lead to any defect. It is desirable not to allow porcelain to come in direct contact with a hard metal screw thread. Normally cement is used between metal and the porcelain. It is seen that cement so used does not cause fracture by extension or contraction.

6.4 Types of Insulators

Over head line insulators are classified as 4 types. They are,

1. Pin type Insulator
2. Suspension type Insulator
3. Strain type Insulator
4. Shackle type Insulator

6.4.1 Pin Type Insulator

As the name suggest, the pin type insulator is designed to be mounted on a pin which in turn is installed on the cross-arm of the pole. The insulator is screwed on the pin and electrical conductor is placed in the groove at the top of the insulator and is tied down with soft copper or aluminum binding wire according to material of the conductor. Pin type insulator is shown in below fig.

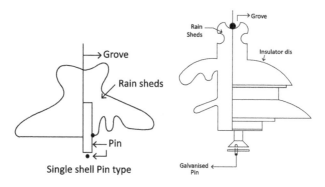

Fig. 6.1

Pin type insulators can be used for the operating voltage up to 33KV. It may be permissible for up to 50KV. Beyond which this insulator is not economical. If the operating voltage increases beyond 33KV, the size of the insulator has to be increased to provide necessary insulation. Due to increase in size, the insulator will become bulky. Also the cost of the insulator will crease with increase in size and operating voltage. Cost of the pin type insulator will vary as the relation given below.

$$\text{Cost} \propto V^2,$$

where $x > 2$

So Cost $\propto V^2$. Hence pin type insulators can not be used for higher operating voltages.

6.4.2 Suspension Type Insulators

For the higher voltages (> 50 KV), pin type insulator becomes bulky and uneconomical. In this case suspension type insulators are used to insulate the high voltage transmission lines.

The suspension insulator hangs from the cross-arm and the line conductor is connected to the bottom of the insulator. These insulators consist of a no. of discs connected in series by metal links in the form of a "string". Each disc or unit in the string is designed for 11 KV. The no. of discs in the string of suspension insulator depends upon the working voltage, size of the insulator and weather conditions. This insulator is arranged in vertical position.

The following are the advantages of suspension type insulator over the pin type insulator,

1. As suspension insulator consists of number of discs, it can be used for high voltage applications.
2. in case of increase or decrease in the operating voltage of the string. The number of discs can be added or removed easily.
3. In case of failure of any one disc, the complete string need not be removed. Only the damaged string can be removed with the other units remaining in the operation.
4. It is more reliable because damage to any one disc will not interrupt the operation of other discs.
5. This type of insulator provides good flexibility to the conductor to which it is connected and here the tension on the conductor is reduced.
6. For the same operating voltage, suspension type insulators are cheaper when compared to pin type insulators.
7. As the conductor is hung below the cross-arm, the tower functions as a lightning rod, this arrangement provides partial protection from lightning.

The construction of suspension the insulator is shown below:

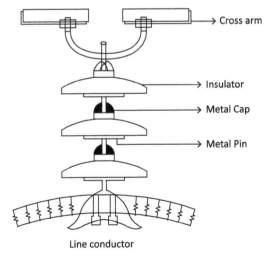

Fig. 6.2

6.4.3 Strain Type Insulator: (Or) Tension Insulator

Strain type insulators are used when the line is subjected to greater tension such as dead ends, river crossings, sharp curves and when there is a change in the direction of line. The main function of this insulator is to reduce excessive tension on the line. It is basically an assembly of suspension insulator used in horizontal position as shown in below fig.

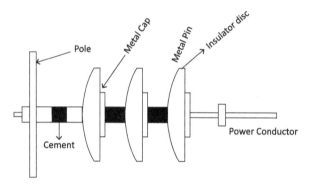

Fig. 6.3

A strain insulator is designed with considerable good strength and with necessary dielectric properties. In case, when the tension is exceedingly high, two or more strings of insulator can be used in parallel. For voltages up to 11 KV, shackle type insulators can be used. But for higher voltages, strain type insulators should be used.

6.4.4 Shackle Type Insulator

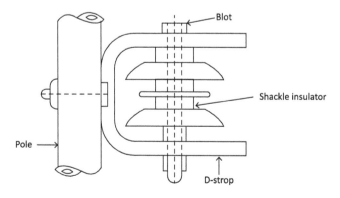

Fig. 6.4

The above figure shows a shackle type insulator. Shackle insulator mounted in a clamp. These insulators can be directly fixed to the pole with a bolt or to the cross-arm. The conductor in the groove is fixed with a soft binding wire. These types of insulators are used for low voltage lines at dead ends, lake crossings and street ends. These can be used for distributed lines.

6.5 Voltage Distribution Among Discs (Units) of Suspension Type Insulator

The above figure represents that the suspension type insulator consists of 4 discs (units) connected together and formed as a string of Insulators. The power conductor is connected to the bottom of the disc of the string and top disc is connected to cross-arm of the tower.

It is found that the voltage impressed on a string of suspension insulators (the voltage applied between the line conductor and earth) does not distribute itself uniformly across the individual discs. The disc nearer to the conductor experiences more voltage than disc near to cross-arms. The inequality of voltage distribution between individual units is all the more pronounced with a larger number of insulator units.

Each string insulator behaves itself like a capacitor having a dielectric medium between two metallic plates (i.e., pin & cap). The capacitance due to metal fittings on the side of an insulator is known as "Self (or) Mutual Capacitance". Further there is also a capacitance between metal fitting of each unit and earthed tower called "Shunt capacitance".

If the distance between tower and string is more, then the effect of shunt capacitance is less, hence charging current is less. So the current at each unit is almost same, hence there is no question of unequal voltage distribution. But this is will not happened in practice. So shunt capacitance plays a vital role in voltage distribution. Due to shunt capacitance, the voltage distribution across each disc will not be the same. Voltage distribution and current of each disc is shown in below fig.

From the figure:

Shunt and self capacitances be expressed with the help of multiplication factor 'K'.

$$\therefore K = \frac{c^|}{c}$$

$$c^| = KC$$

Applying KCL at node (1),

$$I_2 = I_1 + I_1^|$$

$$WCV_2 = WCV_1 + WC^|V_1$$

$$CV_2 = CV_1 + KCV_1$$

$$V_2 = V_1(1+K)$$

Appling KCL at node (2),

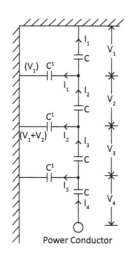

Fig. 6.5

$$I_3 = I_2 + I_2^|$$
$$V_3 WC = V_2 WC + (V_1+V_2)WC^|$$
$$= V_2 WC + (V_1+V_1(1+K))WKC$$
$$V_3 = V_2 + V_1 K + V_1 (1+K)K$$
$$= V_2 + V_1 (K+K+K^2)$$
$$= V_1 (1+K) + V_1 (K^2+2K)$$
$$V_3 = V_1 (K^2+3K+1)$$

Appling KCL at node (3),

$$I_4 = I_3 + I_3'$$
$$V_4 WC = V_3 WC + (V_1 + V_2 + V_3) WC^l$$
$$V_4 WC = V_1(K^2 + 3K + 1) WC(V1 + V_1(1+K))$$
$$+ V_1(K^2 + 3K + 1)) WKC$$
$$= V_1(K^2 + 3K + 1 + K + K + K^2 + K^3 + 3K^2 + K)$$
$$V_4 = V_1(K^3 + 5K^2 + 6K + 1)$$

If number of discs are 5 in the string, then

$$V_5 = V_1(K^4 + 7K^3 + 15K^2 + 10K + 1)$$

String Efficiency:

String efficiency tells us how effective the insulator is utilized in the string and it is defined as the ratio of voltage across the whole string to the product of number of discs and voltage across the disc which is connected to the conductor". It can be expressed as,

$$\% \eta = \frac{\text{Voltage across the string}}{n \times \text{Voltage across the disc which is connected to the conductor}} \times 100$$

(Or)

$$\% \eta = \frac{\text{Spark over voltage across the string}}{n \times \text{Spark over voltage of one disc}} \times 100$$

6.6 Methods to Improve the String Efficiency

The voltage distribution across an insulator string is not uniform due to shunt capacitance effect. The insulator unit nearest to the line conductor is highly stressed while that nearest to the tower is lightly stressed. The drawback of this non-uniform voltage distribution across the units is,

1. If the voltage across the insulator unit nearest to the line conductor reaches beyond a prescribed safe value, it may breakdown and the breakdown of other insulator unit will take place in succession.

2. If the voltage across the insulator unit nearest to the cross-corn is too low, that unit is not utilized to its optimum capacity. Resulting in poor efficiency of the insulator string.

Hence, it is necessary to equalize the voltage distribution across each disc of the string and thereby improving the string efficiency.

There are three methods by which string efficiency can be improved. They are,

1. Using longer cross – arm
2. by capacitance grading
3. by static shielding/Guard ring

6.6.1 Using Longer Cross – Arm

The ratio shunt capacitance to mutual capacitance (K) can be reduced by using longer cross-arm so that the horizontal distance between steel tower to insulator disc is increased there by shunt capacitance will be decreased. As we know that

$$K = \frac{C^|}{C}$$

As shunt capacitance decreases, the value of K is also decreases. If K decreases the voltage across the lowest disc will reduce by which lowest disc experience the less electrical stress. There by string efficiency will be improved. But there is a restriction to increase the length of cross – arm due to high cost and low mechanical strength. So, it has been found that in practice it is not possible to obtain the value of K less than 0.1.

6.6.2 Capacitance Grading

Non – uniform distribution of voltage across an insulator string is due to leakage current from the insulator pin to the supporting structure. This current cannot be eliminated. However, it is possible that discs of different capacities are used such that the product of their capacitive reactance and the current flowing through the respective unit is same. This can be achieved by grading the mutual capacitance of the insulator units. i.e., by having lower units of more capacitance – maximum at the line unit and minimum at the top unit, nearest to the cross – arm.

Consider a 4 unit string. Let C be the capacitance of the top unit and let the capacitances other units be C_2, C_3 and C_4 as shown in fig. As we know,

$C^| = KC$

Applying KCL at node (1),

$$I_2 = I_1 + I_1^|$$

$$V_2 WC_2 = V_1 WC + V_1 WC^|$$

$$VWC_2 = VWC + VWC^|$$

$$[\because V_1 = V_2 = V]$$

$$C_2 = C + KC$$

$$C_2 = C(1+K)$$

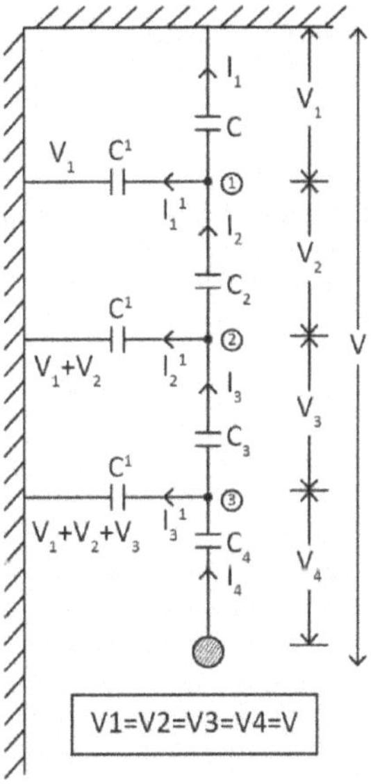

At node (2),

$$I_3 = I_2 + I_2^|$$

$$V_3 WC_3 = V_2 WC_2 + (V_1 + V_2) WC^|$$

$$VWC_3 = VWC_2 + 2\ VWKC$$

$$C_3 = C(1+K) + 2\ KC$$

$$C_3 = C(3K+1)$$

At node (3),

$$I_4 = I_3 + I_3^|$$

$$V_4 WC_4 = V_3 WC_3 + (V_1 + V_2 + V_3) WC^|$$

$$VWC_4 = VWC_3 + 3\ VWKC$$

$$C_4 = C_3 + 3\ KC$$

$$= C(3K+1) + 3KC$$

$$C_4 = C(6K+1)$$

Fig. 6.6

In general, for 'n' discs,

$$C_n = C_{n-1} + (n-1)C^|$$

6.6.3 Static Shielding / Guard Ring Method

In this method, a guard ring surrounding the bottom disc (or unit) and electrically connected to the metal work at the bottom of this unit and therefore to the line conductors. By this the capacitance is created between disc pin to guard ring. Consequently, pin to supporting structure charging currents are exactly cancelled so that same current flows through the identical insulator units and produce equal voltage drops across each insulator unit. The capacitance created by guard ring is Called as "Static capacitance".

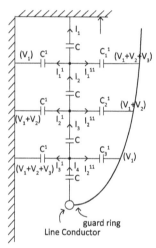

Fig. 6.7

Calculation of Static Capacitance:

At Node 1:

$$I_1^{||} = I_1^{|}$$

$$(V_3 + V_2 + V_1)WC_1^{|} = V_1 WC^{|}$$

$$3\ VWC_1^{|} = VWC^{|}$$

$$3\ C_1^{|} = C^{|}$$

$$C_1^{|} = \frac{C^{|}}{3}$$

At Node 2:
$$I_2^{||} = I_2^{|}$$
$$(V_2 + V_1)WC_2^{|} = (V_1 + V_2)\ WC^{|}$$
$$2\ VWC_2^{|} = 2\ VWC^{|}$$
$$C_2^{|} = C^{|}$$

At Node 3:
$$I_3^{||} = I_3^{|}$$
$$V_4 WC_3^{|} = (V_1 + V_2 + V_3)\ WC^{|}$$
$$VWC_3^{|} = 3\ VWC^{|}$$
$$C_3^{|} = 3\ C^{|}$$

In general, the static capacitance can be written for m^{th} disc is

$$C_m^{|} = \frac{mC^{|}}{n - m}$$

Where $m \neq n$ and 'n' is no. Of discs and m is a perticular disc.

1. **What is the string efficiency of the system shown in fig. If shunt capacitance is half of self capacitance.**

Sol: Given data:

Shunt capacitance $= \dfrac{1}{2}$ self capacitance

$$C^{|} = \frac{1}{2}C$$

$$K = \frac{C^{|}}{C} = 0.5$$

Voltage across disc 2, $V_2 = V_1(1+K)$

$$= V_1(1+0.5) = 1.5\ V_1$$

Voltage across string $= V_1 + V_2$
$$= V_1 + 1.5\ V_1$$
$$= 2.5\ V_1$$

String efficiency % $\eta = \dfrac{V}{n \times V_2} \times 100$

$$= \dfrac{V_1 + V_2}{2 \times V_2} \times 100 = \dfrac{V_1 + 2.5\ V_1}{2 \times 2.5\ V_1} \times 100$$

$$= \dfrac{3.5\ V_1}{5\ V_1} \times 100$$

$$= 83.3\ \%$$

2. A 3ϕ, 3 wire $33\ KV$ system is having a string of 3 discs, the shunt capacitance is $\dfrac{1}{10}$ th self capacitance, what are the voltage drop across each disc.

Sol: Given data:

$$C_1 = \dfrac{1}{10} C \Rightarrow \dfrac{C^1}{C} = \dfrac{1}{10} \Rightarrow K = 0.1$$

Operating voltage $= 33\ KV$ (line voltage)

Phase voltage $= \dfrac{33}{\sqrt{3}}$

Voltage of disc – 1 is V_1

Voltage of disc – 2 is $V_2 = V_1 (1 + K)$
$$= V_1 (1 + 0.1)$$
$$V_2 = 1.1\ V_1$$

Voltage of disc – 3, $V_3 = V_1 (K^2 + 3K + 1)$
$$= V_1 \left((0.1)^2 + 3(0.1) + 1 \right)$$
$$= V_1 (1.31)$$
$$V_3 = 1.31\ V_1$$

Voltage across the string $V = V_1 + V_2 + V_3$

$$= V_1 + 1.1\ V_1 + 1.3\ V_1$$

$$\frac{33}{\sqrt{3}} = 3.4\ V_1$$

$$V_1 = \frac{33}{\sqrt{3}} \times 3.4$$

$$= 5.6\ KV$$

$$V_2 = 5.6\ (1+0.1)$$

$$V_2 = 6.7\ KV$$

$$V_3 = 5.6\ (1.31\ V_1)$$

$$V_3 = 7.3\ KV$$

3. A 3 – φ, 33 KV transmssion line is connected to a 3 – discs suspension type insulator. If the shunt capacitance is 11% of the self capaciance. Find (a) voltage across each disc of the string (b) string efficiency

A. Phase voltage $V = \dfrac{33}{\sqrt{3}} = 19.05\ KV$

Shunt capacitance $C^| = 0.11 \times C$

$$\frac{C^|}{C} = 0.11 = K$$

$$V_1 = V_1$$

$$V_2 = V_1(1+K) = V_1(1+0.11) = 1.11\ V_1$$

$$V_3 = V_1(K^2 + 3K + 1) = V_1(0.11^2 + 3(0.11) + 1) = 1.342\ V_1$$

Total voltage across the string

$$V_1 + V_2 + V_3 = V_{Ph}$$

$$V_1 + 1.11\ V_1 + 1.342\ V_1 = 19.05\ KV$$

$$V_1 = 5.51\ KV$$

$$V_2 = 1.11\ V_1 = 1.11 \times 5.51 = 6.125\ KV$$

$$V_3 = 1.342\ V_1 = 1.342 \times 1.11 = 7.39\ KV$$

String efficiency $\eta = \dfrac{19.05}{3 \times 7.39} \times 100 = 85.87\ \%$

4. A string of four insulator disc connected to a 285 KV line. If the self capacitance is 5 times more pin to earth capacitance. Calculate (a) potential difference across each unit (b) string efficiency.

A. Phase voltage $V_{Ph} = \dfrac{285}{\sqrt{3}} = 164.54\ KV$

Self capacitance $C = 5 \times C^l$

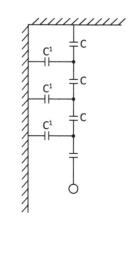

$\dfrac{C^l}{C} = \dfrac{1}{5} = 0.2$

$V_1 = V_1$

$V_2 = V_1(1+K) = V_1(1+0.2) = 1.2\ V_1$

$V_3 = V_1\left(K^3 + 5K^2 + 6K + 1\right)$

$= V_1\left((0.2)^3 + 5(0.2)^2 + 6(0.2) + 1\right)$

$= 2.4\ V_1$

Total voltage across the string

$V_1 + V_2 + V_3 + V_4 = 164.54\ KV$

$V_1 + 1.2\ V_1 + 1.64\ V_1 + 2.4\ V_1 = 164.54\ KV$

$V_1 = 26.36\ KV$

$V_2 = 1.2\ V_1 = 1.2 \times 26.36 = 31.63\ KV$

$V_3 = 1.64\ V_1 = 1.64 \times 26.36 = 43.23\ KV$

$V_4 = 2.4\ V_1 = 2.4 \times 26.36 = 63.26\ KV$

String efficiency $= \dfrac{164.54}{4 \times 63.26} \times 100 = 65.02\ \%$

5. A 3-ϕ transmission line is connected to a suspension type insulation which is having 5 insulator discs. The self capacitance is 5 times the disc to tower capacitance calculate (a) voltage across each unit percentage of total voltage of the string (b) string efficiency.

A. Total number of disc = 5

Self capacitance $C = 5\, C^l$

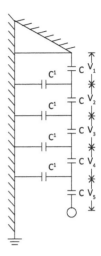

$$K = \frac{C^l}{C} = \frac{1}{5} = 0.2$$

$V_1 = V_1$

$V_2 = V_1(1+K) = 1.2\, V_1$

$V_3 = V_1(K^2 + 3K + 1) = 1.64\, V_1$

$V_4 = V_1(K_3 + 5K^2 + 6K + 1) = 2.40\, V_1$

$V_5 = V_1(K^4 + 7K^3 + 15K^2 + 10K + 1) = 3.65\, V_1$

Total voltage across the string
$V = V_1 + V_2 + V_3 + V_4 + V_5$

$V = 1.2\, V_1 + 1.64\, V_1 + 2.40\, V_1 + 3.65\, V_1 = 9.89\, V_1$

$V_1 = \dfrac{V}{9.89} \times 100 = 10.11\%$ Of V

$V_2 = 1.2\, V_1 = 12.13\%$ of V

$V_3 = 1.64\, V_1 = 16.58\%$ of V

$V_4 = 2.40\, V_1 = 24.26\%$ of V

$V_5 = 3.65\, V_1 = 36.90\%$ of V

String efficiency $\eta = \dfrac{V}{5 \times 0.369} \times 100 = 54.2\%$

6. A four insulator disc of suspension type insulator is connected to a $3-\phi$, $66\, KV$ line. If the shunt capacitance is one tenth of self capacitance. Find the voltage across fourth disc and also calculate string efficiency.

A. Number of insulators = 4

Self capacitance $C = 10 \, C^l$

$$K = \frac{C^l}{C} = 0.1$$

$V_1 = V_1$

$V_2 = V_1(1+K) = 1.1 \, V_1$

$V_3 = V_1(K^2 + 3K + 1) = 1.31 \, V_1$

$V_4 = V_1(K^3 + 5K^2 + 6K + 1) = 1.65 \, V_1$

Total voltage across the string

$$\frac{66}{\sqrt{3}} = V_1 + V_2 + V_3 + V_4$$

$38.10 \, KV = V_1 + 1.1 \, V_1 + 1.31 \, V_1 + 1.65 \, V_1$

$V_1 = 7.52 \, KV$

Voltage across fourth disc $V_4 = 1.65 \, V_1 = 1.65 \times 7.52 = 12.4 \, KV$

String efficiency $\eta = \dfrac{38.10}{4 \times 12.4} \times 100 = 76.8\%$

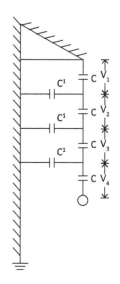

7. **A 3-φ overhead transmission line is suspended by a suspension type insulator which consists of 3 units. The potential across top unit and middle unit are $7 \, KV$ and $10 \, KV$ respectively. Calculate the (i) Ratio of capacitance between pin and earth to self capacitance of each unit (ii) Line voltage (c) String efficiency**

A. Given data:

$V_1 = 7 \, KV$
$V_2 = 10 \, KV$

$V_2 = (1+K)V_1$

$10 = (1+K)7$

$K = \dfrac{3}{7} = 0.428$

Mechanical Design of Overhead Transmission Lines | 195

The ratio of pin to earth capacitance to the self capacitance = 0.428

Self capacitance = 0.428

$V_3 = V_1(K^2 + 3K + 1) = 17.05\ KV$

Total voltage across the string
$= V_1 + V_2 + V_3 = 34.27\ KV$

Line voltage $= \sqrt{3}(V_1 + V_2 + V_3) = 58.98\ KV$

String efficiency $\eta = \dfrac{34.27}{3 \times 17.05} \times 100 = 66.14\%$

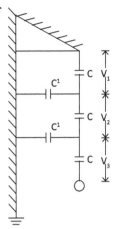

8. **An insulator string containing 5 units has equal voltage across each unit by using disc of different capacitances. If the top unit has a capacitance of C and pin to tower capacitance of all units is 20% of mutual capacitance of top unit. Calculate mutual capacitance of each disc.**

A. Given data:

Pin to tower capacitance $C^| = 0.2\ C$

$K = \dfrac{C^|}{C} = 0.2$

$C_1 = C$

$C_2 = C_1(1+K) = C(1+0.2) = 1.2\ C$

$C_3 = C_1(1+3K) = C(1+(0.2)3) = 1.6\ C$

$C_4 = C_1(1+6K) = C(1+6(0.2)) = 2.2\ C$

$C_5 = C_1(1+10K) = C(1+10(0.2)) = 3C$

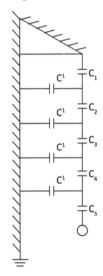

9. **A 6 insulator disc is connected to a $3-\phi$ transmission line the shunt capacitance of each disc disc is $C^|$ and the conductor is surrounded by the guard ring calculate the capacitance b/w pin to line of each disc of the string.**

$$C_m^l = \frac{mC^l}{n-m}$$

$$C_1^l = \frac{C^l}{6-1} = \frac{C^l}{5}$$

$$C_2^l = \frac{2C^l}{4} = \frac{C^l}{2}$$

$$C_3^l = \frac{3C^l}{3} = C^l$$

$$C_4^l = \frac{4C^l}{2} = 2\,C^l$$

$$C_5^l = \frac{5C^l}{1} = 5\,C^l$$

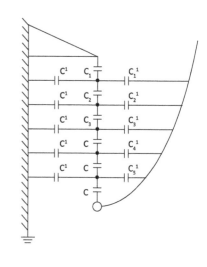

10. Each of 3 insulators forming a string has a self capacitance of C farad. The shunt capacitance of each insulator is $0.2C$ to earth and $0.1C$ to line. A guard ring increases the capacitance of the line of metal work of lowest insulator disc to $0.3C$. Calculate the string efficiency of the string with and without guard ring.

A. (a) Without considering the effect of guard ring

Applying KCL at node 1

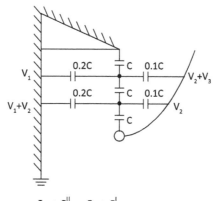

$$I_2 + I_1^{||} = I_1 + I_1^{|}$$

$$WCV_2 + (V_2 + V_3)W\,0.1\,C = V_1 WC + V_1 W\,0.2\,C$$

$$1.1\ V_2 + 0.1\ V_3 = 1.2\ V_1$$
$$V_3 = 12\ V_1 - 11\ V_2 \quad \rightarrow (1)$$

Appling KCl at node (2)
$$I_3 + I_2^{\|} = I_2 + I_2^{\prime}$$
$$WCV_3 + W\,0.1\ CV_3 = (V_1 + V_2)W\,0.2\ C + WCV_2$$
$$V_3 = 0.18\ V_1 + 1.09\ V_2 \quad \rightarrow (2)$$

From (1) and (2)
$$12\ V_1 - 11\ V_2 = 0.18\ V_1 + 1.09\ V_2$$
$$V_2 = 0.97\ V_1$$
$$V_3 = 12\ V_1 - 11\ V_2$$
$$= 12\ V_1 - 11(0.97\ V_1)$$
$$= 1.33\ V_1$$

Total voltage across the string $= V_1 + V_2 + V_3 = 3.3\ V_1$

String efficiency $\eta = \dfrac{3.3\ V_1}{3 \times 1.33\ V_1} \times 100 = 82.70\%$

(b) Considering the effect of guard ring

Applying KCl at node 2
$$I_3 + I_2^{\|} = I_2 + I_2^{\prime}$$
$$WCV_3 + W\,0.3\ CV_3 = W\,0.2\ C(V_1 + V_2) + WCV_2$$
$$1.3\ V_3 = 0.2\ V_1 + 1.2\ V_2$$
$$V_3 = 0.153\ V_1 + 0.92\ V_2 \quad \rightarrow (3)$$

From (1) and (3)
$$12\ V_1 - 11\ V_2 = 0.153\ V_1 + 0.92\ V_2$$
$$11.84\ V_1 = 11.92\ V_2$$
$$V_2 = 0.99\ V_1$$

$$V_3 = 0.153\ V_1 + 0.92\ V_2$$
$$= 0.153\ V_1 + 0.92(0.99\ V_1) = 1.070\ V_1$$

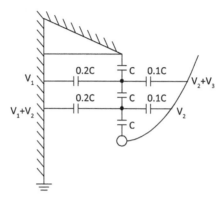

Total voltage across the string $= V_1 + V_2 + V_3 = 3.109\ V_1$

String efficiency $= \dfrac{3.109\ V_1}{3 \times 1.070\ V_1} \times 100 = 95.07\%$

11. **The self capacitance of each unit in a string of 3 suspension type insulator disc. The shunting capacitance of each insulator to earth $0.5C$ while for line $0.1C$. Calculate voltage across each insulator and string efficiency.**

A. Given data:

Pin to earth capacitance $= 0.5\ C$

Pin to line capacitance $= 0.1\ C$

Applying KCl at node 1

$$I_2 + I_1^{||} = I_1 + I_1^{|}$$

$$W C V_2 + W\,0.1\,C(V_2 + V_3) = W\,0.5\,CV_1 + WV_1 C$$

$$V_2 + (V_3 + V_2)0.1 = 0.5\ V_1 + V_1$$

$$1.1\ V_2 = 1.5\ V_1 - 0.1\ V_3$$

$$V_2 = 1.36\ V_1 - 0.09\ V_3 \qquad \rightarrow (1)$$

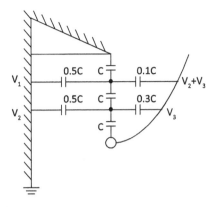

Applying KCl at node 2

$$I_3 + I_2'' = I_2 + I_2'$$

$$WCV_3 + W\,0.1\,CV_3 = WCV_2 + W\,0.5\,C(V_1 + V_2)$$

$$V_3 + 0.1\,V_3 = V_2 + 0.5\,V_1 + 0.5\,V_2$$

$$V_2 = 0.733\,V_3 - 0.33\,V_1 \qquad \rightarrow (2)$$

From (1) and (2)

$$1.36\,V_1 - 0.09\,V_3 = 0.733\,V_3 - 0.33\,V_1$$

$$V_3 = 2.05\,V_1$$

$$V_2 = 0.733\,(2.05\,V_1) - 0.33\,V_1$$

$$= 1.17\,V_1$$

String efficiency $= \dfrac{4.22\,V_1}{3 \times 2.05\,V_1} \times 100 = 68.6\%$

6.7 Sag & Tension Calculations

Overhead lines are supported on mechanical structures consisting of components like insulators, cross – arms, poles or towers etc. The strength of these components must be such that there is no mechanical failure of line, even under the worst weather conditions.

The conductor is acted upon the forces such as weight of the conductor itself, wind pressure and tension. A conductor stretched between two supports will have an ultimate strength at which it will fail and the ultimate strength of a conductor depends upon the type of conductor material used for overhead line.

While stringing overhead lines, it is necessary to allow a reasonable factor of safety in respect of the tension to which the conductor is subjected. The tension in the conductor is normally expected to be less than 50% of its tensile strength. The tension in a conductor depends on the diameter of the conductor, length of the conductor between supports, material of conductor, and sag in conductor, wind pressure and temperature. The relationship between tension and sag is dependent on the loading conditions and temperature variations.

Sag: "The difference in level between the points of supports and lowest point on the conductor is known as Sag".

6.8 Factors Affecting Sag

The following are the factors which affects the sag,

1. **Weight of the conductor:**

    This affects the sag directly. Heavier the conductor, greater will be the sag. In locations, where ice formation takes place on the conductor, this will also cause increase in the sag. So sag is directly proportional to weight of the conductor.

2. **Length of the span:**

    This also affects the sag directly. Sag is directly proportional to the square of the span length. Hence other conditions, such as type of conductor, working tension, temperature etc., remaining the same. A section with longer span will have much greater sag.

3. **Working Tensile Strength:**

    The sag is inversely proportional to the working tensile strength of conductors if other conditions such as temperature, length of span remain the same. Working tensile strength of the conductor is determined as follows,

Working Tensile strength (or) Tension

$$= \frac{Ultimate Stress \times Area of cross section}{Factor of safety}$$

$$T = \frac{Tensile sterss \times area of cross section}{Factor of safety}$$

4. **Temperature:**

All metallic bodies expand with the rise in temperature and therefore the length of the conductor increases this in turn increase the sag. So sag is directly proportional to temperature variations.

The sag plays an important role in the design of over head line. It is disadvantageous to provide either too high sag or too low sag. In case the sag is too high, more conductor material is required this will cause to increase the cost of the conductor, more weight on the supports is to be supported, higher support are necessary and there is a chance of greater swing – amplitude due to wind load. On the other hand in the case of too low sag, there is more tension in the conductor and thus the conductor is liable to break if any additional stress is to be taken, such as stress due to vibration of line or stress due to fall in temperature.

6.9 Calculation of Sag & Tension

6.9.1 With Equal Level Supports

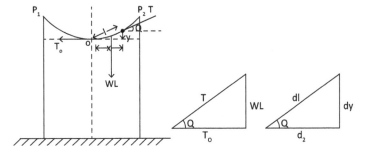

Fig. 6.8

Let the conductor be suspended between points P_1 and P_2 with 'O' as the lowest point of the conductor as shown in fig (a). Let 'P' be a point on the conductor, draw a tangent at a point 'P' as shown in fig (a).

Consider a position 'OP' of a curve length 'l' of conductor hanging in air with 'O' as the lowest point on the conductor. Let the weight of the conductor per meter length be 'W' kgs.

The portion 'OP' in equilibrium under the action of three forces and they are,

1. $T_0 \rightarrow$ Horizontal Tension at 'O'
2. $T \rightarrow$ Tangential Tension at 'P'
3. $WL \rightarrow$ Weight of conductor acting vertically downwards through the centre of gravity

Above three forces can be represented by a triangle as shown in fig (b) and from this,

$$\tan\theta = \frac{Wl}{T_0} \quad \rightarrow (1)$$

The horizontal and vertical distances of the length 'OP' of the conductor are x, y respectively. Since the co-ordinates of the point 'P' are (x, y). From fig (c),

$$\tan\theta = \frac{dy}{dx} \quad \rightarrow (2)$$

And also $dl^2 = dx^2 + dy^2$

$$\left(\frac{dl}{dx}\right)^2 = 1 + \left(\frac{dy}{dx}\right)^2$$

$$= 1 + \tan^2\theta$$

$$\left(\frac{dl}{dx}\right) = \sqrt{1 + \left(\frac{Wl}{T_0}\right)^2}$$

$$dx = \frac{dl}{\sqrt{1 + \left(\frac{Wl}{T_0}\right)^2}}$$

Integrating the above equation on both sides

$$x = \frac{T_0}{W} \sinh^{-1} \frac{Wl}{T_0} + c$$

Where 'c' is the integration constant

From initial conditions, $x = 0$ & $l = 0$ then we get $c = 0$

∴ Above equation becomes,

$$x = \frac{T_0}{W} \sinh^{-1} \frac{Wl}{T_0} \quad \rightarrow (3)$$

$$\frac{Wl}{T_0} = \sinh\left(\frac{Wx}{T_0}\right)$$

$$Wl = T_0 \sinh\left(\frac{Wx}{T_0}\right) \quad \rightarrow (4)$$

$$l = \frac{T_0}{W} \sinh\left(\frac{Wx}{T_0}\right) \quad \rightarrow (5)$$

We know that, $\dfrac{dy}{dx} = \tan\theta = \dfrac{Wl}{T_0}$

$$= \frac{W}{T_0} \times \frac{T_0}{W} \sinh\left(\frac{Wx}{T_0}\right)$$

$$dy = \sinh\left(\frac{Wx}{T_0}\right) \cdot dx \quad \rightarrow (6)$$

Integrating the above equation on both sides,

$$y = \frac{T_0}{W} \cosh\left(\frac{Wx}{T_0}\right) + D \quad \rightarrow (7)$$

Where 'D' is an integrating constant

From initial conditions,

$x = 0$ & $y = 0$, then we get $D = -\dfrac{T_0}{W}$

Equation (7) becomes,

$$y = \frac{T_0}{W}\cosh\left(\frac{Wx}{T_0}\right) - \frac{T_0}{W}$$

$$y = \frac{T_0}{W}\left[\cosh\left(\frac{Wx}{T_0}\right) - 1\right]$$

Above equation represents a curve and is known as "Catenaries Curve".

Now,

$$\cosh\left(\frac{Wx}{T_0}\right) = 1 + \frac{\left(\frac{Wx}{T_0}\right)^2}{2!} + \frac{\left(\frac{Wx}{T_0}\right)^4}{4!} + \ldots$$

So,

$$y = \frac{T_0}{W}\left[1 + \frac{W^2 x^2}{2\,T_0^2} + \frac{W^4 x^4}{24\,T^4} - 1\right]$$

$$y = \frac{T_0}{W}\left(\frac{W^2 x^2}{2\,T_0^2} + \frac{W^4 x^4}{24\,T^4}\right)$$

If 4th and higher order terms are neglected, then

$$y = \frac{T_0}{W} \times \frac{W^2 x^2}{2T_0^2}$$

$$y = \frac{Wx^2}{2\,T_0} \qquad \rightarrow (8)$$

From fig (b), the tension T at point 'P' is given by

$$T = \sqrt{T_0^2 + (Wl)^2}$$

$$= \sqrt{T_0^2 + \left(T_0 \sinh\left(\frac{Wx}{T_0}\right)\right)^2} \quad \text{From equation (4)}$$

$$= \sqrt{T_0^2\left(1 + \sinh^2\left(\frac{Wx}{T_0}\right)\right)}$$

$$T = T_0\sqrt{\cosh^2\left(\frac{Wx}{T_0}\right)}$$

$$T = T_0 \cosh\left(\frac{Wx}{T_0}\right)$$

If the line is supported between two poles P_1 and P_2 at the same Level than the length of span is 'L', then at the supports

$$x = \frac{L}{2}, \text{ Then}$$

$$T = T_0 \cosh\left(\frac{WL}{2T_0}\right)$$

Sag 'S' is the value of 'y' at P_1 or P_2 is given by

$$S = \frac{W\left(\frac{1}{2}\right)^2}{2T_0}$$

$$S = \frac{WL^2}{8T_0}$$

Length of the line in a half span $L = \dfrac{T_0}{W}\sinh\left(\dfrac{WL}{2T_0}\right)$ from eq (5)

$$= \frac{T_0}{W}\left[\frac{WL}{2T_0} + \frac{W^3 L^3}{48 T_0^3}\right]$$

Neglecting terms of order exceeding cube (3rd) order,
We have length of the line in half span is,

$$= \frac{T_0}{W} \times \frac{WL}{2T_0} + \frac{T_0}{W} \times \frac{W^3 L^3}{48 T_0^3}$$

$$= \frac{L}{2} + \frac{W^2 L^3}{48 T_0^2}$$

$\therefore$ Length of the line in half span $= \dfrac{L}{2}\left[1 + \dfrac{W^2 L^2}{24 T_0^2}\right]$

Length of line in full span $= 2\left[\dfrac{L}{2} + \dfrac{W^2 L^3}{48\, T_0^2}\right]$

$= L + \dfrac{W^2 L^3}{24\, T_0^2}$

Length of the line in full span $= L\left[1 + \dfrac{W^2 L^2}{24\, T_0}\right]$

6.9.2 At Unequal Level of Supports

When transmission lines are run on steep inclines as in case of hilly areas, the two supports A and B will be at different levels. The shape of the conductor strung between the supports may be assumed to be a part of the parabola. In this case, the lowest point of the conductor will not lie in the middle of the span.

Consider two poles with different heights and the conductor is stretched between these two towers P_1 & P_2. The sag of tower P_1 w.r.t to conductor is 'S_1', and for tower P_2 is S_2. The distance of pole P_1 from the tower point of conductor is x_1 and tower P_2 is x_2. Let the difference in levels between the two supports be 'h' and the lowest point of the conductors be 'O'.

Span length $= L$

Difference in levels between two supports $h = S_2 - S_1$

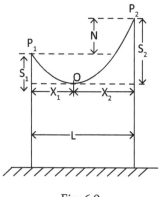

Fig. 6.9

Tension in the conductors $= T$

If 'W' is weight per unit length of conductor, then the sag $S_1 = \dfrac{W x_1^2}{2T}$

And sag $S_2 = \dfrac{Wx_2^2}{2T}$

$x_1 + x_2 = L$ → (1)

Now $S_2 - S_1 = \dfrac{Wx_2^2}{2T} - \dfrac{Wx_1^2}{2T}$

$= \dfrac{W}{2T}\left(x_2^2 - x_1^2\right)$

$= \dfrac{W}{2T}(x_2 + x_1)(x_2 - x_1)$

$S_2 - S_1 = \dfrac{WL}{2T}(x_2 - x_1)$

$h = \dfrac{WL}{2T}(x_2 - x_1)$

$(x_2 - x_1) = \dfrac{2Th}{WL}$ → (2)

Solving equation (1) & (2),

$x_1 = \dfrac{L}{2} - \dfrac{Th}{WL}$

$x_2 = \dfrac{L}{2} + \dfrac{Th}{WL}$

6.10 Effect of Wind and Ice on the Weight of the Conductor

In areas where it becomes too cold in winter, there is a possibility of formation of an ice coating on the line conductors. The formation of an ice coating on a line conductor has a twofold effect

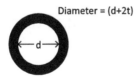

Fig. 6.10

1. It increase the weight of the conductor and
2. It increase the effective diameters of the conductor

In this condition, the weight of conductor, together with weight of ice acts vertically downwards. Thus the total vertical weight acting on the conductor per meter length is $(W_c + W_i)$.

Where $W_c \rightarrow$ Weight of the conductor per meter length

$W_i \rightarrow$ Weight of ice coating per meter length

W Determines as follows:

Let the diameter of conductor be 'd' meters and radial thickness of ice coating be 't' meter as shown in above fig. The over all diameter of ice covered conductor $(d \quad t)$ meter.

Volume of ice coating per meter length of conductor,

$$= \frac{\Pi}{4}\left[(d+2t)^2 - d^2\right]$$

$$= \frac{\Pi}{4}\left[d^2 + 4t^2 + 4dt - d^2\right]$$

$$= \Pi t(d+t) m^3$$

The density of ice is approximately 920 kg/m^3

So, the weight of ice coating per meter length $W_i = 920 \times \Pi t(d+t)$

$$W_i = 920 \times \Pi t(d+t)$$

Due to weight of ice depends on the line and the wind pressure, the mechanical stress increases in the conductor and therefore the line must be designed to withstand these stresses and tensions. Under this condition, the weight of the conductor together with weight of ice acts vertically downwards while the wind loading W_w acts horizontally as shown in below fig.

Resultant weight per meter length of conductor including ice coating and wind force,

$$W_r = \sqrt{(W_c + W_i)^2 + W_w^2}$$

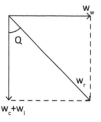

Where W_w = wind force in kg per meter length

= Wind pressure per m^2 of projected area × projected area per meter length

$$W_w = P \times (d + 2t)$$

∴ Maximum sag $= \dfrac{W_r L^2}{8T}$ → this can be called as "slant sag"

When the ice and wind are acting simultaneously, the lowest point of the conductor does not remain vertically down but away from it at an angle 'θ' given by the expression

$$\theta = \cos^{-1} \dfrac{(W_c + W_i)}{W_r}$$

The vertical sag will be obtained by multiplying the slant sag with '$\cos \theta$'.

∴ Vertical sag $= S \times \cos \theta$

Vertical sag = slant sag $\times \cos \theta$

6.11 Sag Template

In the initial planning stages, a survey of the proposed route enables an estimated line profile to be drawn. Such a profile is constructed with the horizontal scale than the vertical scale. This profile should meet the minimum clearance requirement and the location of the supports. The sag template is usually made on celluloid or tracing cloth.

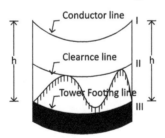

Fig. 6.11

The above figure represents the sag template graph. In which the upper curve I represents the line conductor. The middle curve is below the upper curve I by a uniform vertical distance equal to the desired minimum vertical clearance to ground. This clearance to ground is governed by the operating voltage and according to *IE* rules. The lower curve i.e., curve III is below the middle curve by a uniform vertical distance equal to the height of a standard tower measured to the point of support of the conductor. If the location of the left tower has been decided, the location of the right hand tower can be determined by adjusting the sag template so that the conductor line passes through the point of support on the left hand tower and the clearance line is tangent to ground at one or more points.

Sag template is very convenient method for allocating the positions and height of towers/supports correctly on the profile.

Uses of sag template:

1. The required span length and height of the tower is determined in order to maintain min. ground clearance.
2. Point of errection of towers is achieved.
3. Towers are designed such that they carries equal load.
4. Towers are designed such that the height and weight of the tower is kept constant.
5. Economical layout is achieved.

1. **A Tr. line has a span of $200\,mtrs.$ between level supports, the cross sectional area of conductor is $1.29\,cm^2$ weighs $1170\,kg/km$ and has a breaking stress of $4218\,kg/cm^2$. Calculate the sag for a factor of safety 5, allowing wind pressure $122\,kg/m^2$ of projected area and vertical sag.**

Sol: Given data:

Span length $L = 200\,mtrs$

Cross sectional area $= 1.29\,cm^2$

Weight of conductor $W_c = 1170 \dfrac{Kg}{km} = 1.17\,kg/m$

Breaking stress $= 4218\,kg/cm^2$

Factor of safety $= 5$

Wind pressure $P = 122\,kg/m^2$

$$\text{Tension} = \dfrac{Ulitimate stress\,(or)\,Breaking stress \times Area}{Factor of Safety}$$

$$= \frac{4218 \times 1.29}{5}$$

$$= 1088.24 \ kg$$

Cross section of the conductor $a = \frac{\Pi}{4} d^2$

$$d^2 = \frac{4 \times a}{\Pi}$$

$$d = \sqrt{\frac{4 \times a}{\Pi}}$$

$$= \sqrt{\frac{4 \times 1.29}{\Pi}}$$

$$= 1.28 \ cm$$

Weight of wind forces W_w = wind pressure × projected area of conductor

$$= P \times d$$
$$= 122 \times 1.28 \times 10^{-2}$$
$$= 1.56 \ kg/m$$

Resistant weight $W_r = \sqrt{W_C^2 + W_w^2}$

$$V = \sqrt{(1.17)^2 + (1.56)^2}$$

$$W_r = 1.95 \ kg/m$$

Slant sag $S = \dfrac{W_r L^2}{8 \ T}$

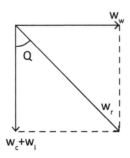

$$= \frac{1.95 \times (200)^2}{8 \times 1088.244}$$

$$= 8959 \text{ mtrs.}$$

Vertical sag $= S \times \cos\theta$

$$\cos\theta = \frac{W_c + W_i}{W_r}$$

$$= \frac{1.17}{1.95}$$

$$= 0.6$$

∴ Vertical sag $= 8.95 \times 0.6$

$$= 5.37 \text{ mtrs}$$

2. An overhead line has the following data, span length $185\,m$, difference in levels of supports $6.5\,m$, conductor dia $1.82\,cm$, weight per unit length of conductor $2.5\,kg/m$, wind pressure $49\,kg/m^2$ of projected area. Max. Tensile stress of the conductor $4250\,kg/m^2$. Factor of safety 5. Calculate available sag in mtrs at the lower support.

Sol: Given data:

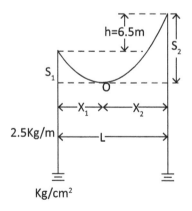

Span length $L = 185\,m$

Difference in level supports $h = 6.5\,m$

Diameter of conductor $d = 1.82$ cm
Weight of the conductor per unit length $W_c = 2.5$ kg/m
Wind pressure $= 49$ kg/m²
Max. Tensile stress of conductor $= 4250$ kg/cm²
Factor of safety $= 5$

Wind force per meter length $W = P \times (d + t)$
$$= P \times d [\because \text{ice} = 0]$$
$$= 49 \times 1.82 \times 10^{-2}$$
$$W_w = 0.89 \text{ kg/m}$$

∴ Resultant weight per meter length $W_r = \sqrt{W_c^2 + W_w^2}$
$$= \sqrt{(2.5)^2 + (0.89)^2}$$
$$W_r = 2.65 \text{ kg/m}$$

Working tension $= \dfrac{\text{Max. tensile stress} \times \text{area of crosssection}}{\text{Factor of safety}}$

$$= \dfrac{4250 \times \dfrac{\Pi}{4} d^2}{5}$$

$$= \dfrac{4250 \times \dfrac{\Pi}{4} (1.82)}{5}$$

$$= \dfrac{11056.59}{5}$$

$$T = 2211.32 \text{ kg}$$

Sag at lower support $S = \dfrac{W_r x_1^2}{2T}$

But $x_1 = \dfrac{L}{2} - \dfrac{Th}{W_r L}$

$$= \dfrac{185}{2} - \dfrac{2211.32 \times 6.5}{2.65 \times 185}$$

$$x_1 = 63.18 \text{ m}$$

$$\therefore S_1 = \frac{2.65 \times (63.18)^2}{2 \times 2211.32}$$

$$= 2.39 \ m$$

3. A transmission line crossing a river is supported from two towers at heights of $30m$ and $80m$ above the water level. The horizontal distance between the towers is $450\ meters$. if the tension in the conductor is $1500\ kg$ and weight of the conductor is $1.4\ kg/m$ length. Find the minimum clearance of the conductor and water and also find clearance mid – way between the supports.

Sol: Given data:

Span length $L = 300m$

Tension $T = 1500\ kg$

Weight of conductor, $W_c = 1.4\ kg/m$

Difference in levels between supports $h = 80 - 30 = 50\ m$

Let x_1, x_2 be the diameters of the lowest point 'O' from the supports A and B respectively.

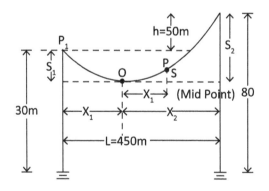

then $x_1 + x_2 = L$

$x_1 + x_2 = 450$ → (1)

$S_1 = \dfrac{Wx_1^2}{2T}$ and $S_2 = \dfrac{Wx_2^2}{2T}$

$$x_2 - x_1 = \frac{2Th}{WL}$$

$$x_2 - x_1 = \frac{2 \times 1500 \times 50}{1.4 \times 50}$$

$$x_2 - x_1 = 238 \text{ m} \qquad \rightarrow (2)$$

$$x_1 = \frac{L}{2} - \frac{Th}{WL} = \frac{400}{2} - \frac{1500 \times 50}{1.4 \times 450}$$

$$= 225 - 119.047$$

$$= 105.9$$

$$\cong 106 \text{ mtrs}$$

$$x_2 = \frac{L}{2} + \frac{Th}{WL} \Rightarrow \frac{450}{2} + \frac{1500 \times 50}{1.4 \times 450}$$

$$= 225 + 119.04$$

$$= 344 \text{ m}$$

$$\therefore S_1 = \frac{1.4 \times (106)^2}{2 \times 1500} \Rightarrow 5.24 \text{ m}$$

Clearance of conductor and water $= 30 - 5.24$

$$= 24.76 \text{ m}$$

Let the midpoint be 'P' at a distance 'x' from the lowest point 'O'. then,

$$x = \frac{l}{2} - x_1$$

$$= \frac{450}{2} - 106$$

$$x = 119 \text{ m}$$

Sag at mid point P, $S = \dfrac{Wx^2}{2T}$

$$= \frac{1.4 \times (119)^2}{2 \times 1500}$$

$$= 6.61 \text{ mtrs.}$$

∴ Clearance of mid point 'P' from water wave $\rho = 24.76 + 6.61$

$$= 31.37 \text{ mtrs.}$$

4. The following data refer to an overhead transmission line laving parabolic configuration: Weight of the conductor per meter length = 1.8 kg/m, cross sectional area is 2.3 cm², ultimate tensile strength = 7800 kg/cm², distance between the support in 500 m, difference of levels of the supports 16m, ice land per meter conductor is 1.1 kg/m. Find the vertical sag from the taller of the two supports, which must be allowed so that factor of safety shall be 4.

Sol: Weight of ice, $W_i = 1.1 \text{ kg/m}$

Weight of conductor, $W_c = 1.8 \text{ Kg/m}$

Area of cross section, $a = 2.3 \text{ cm}^2$

Ultimate strength $= 7800 \text{ Kg/cm}^2$

Span length $= 500 \text{ m}$

Difference of levels of supports $h = 16 \text{ m}$

Factors of safety $= 4$

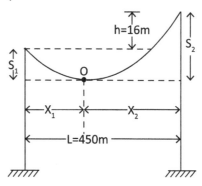

Working Tension $T = \dfrac{7800 \times 2.3}{4} = 4485 \text{ Kgs}$

Resultant weight of the conductor $W_r = W_c + W_i = 1.8 + 1.1 = 2.9 \text{ kg/m}$

$$x_1 + x_2 = 500 \text{ m}$$

$$x_2 = \frac{L}{2} + \frac{Th}{W_r L}$$

$$x_2 = \frac{500}{2} + \frac{4485 \times 16}{2.9 \times 500} = 299.5 \text{ mtrs.}$$

$$x_1 = \frac{500}{2} - \frac{4485 \times 16}{2.9 \times 500}$$

∴ Sag from the taller of two supports $S_2 = \dfrac{W_r x_2^2}{2T}$

$$= \frac{2.9 \times (299.5)^2}{2 \times 4485}$$

$$- 29 \text{ mtrs.}$$

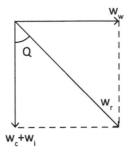

Vertical sag $= S_2 \cos\theta$

$$\cos\theta = \frac{W_c + W_i}{W_r}$$

$$= \frac{2.9}{2.9}$$

$$= 1$$

∴ Vertical sag $= 29 \times 1$

$$= 29$$

5. A transmission line conductor has an effective diameter of 19.5 mm and weight 1.0 kg/m. If the maximum permissible sag with a horizontal wind pressure of 39 kg/m2 of projected area and 12.7 mm

radial ice coating is 6.3 m. calculate the permissible span between two supports at the same level allowing a safety factor of 2. Strength of the conductor is 800 kg and weight of ice is 910 kg/m3.

Sol: Conductor diameter $d = 19.5$ mm

Weight of the conductor $W_c = 1.0$ kg/m

Wind pressure $W_w = 39$ kg/m^2

Thickness of ice $t = 12.7$ mm

Permissible sag $S = 6.3$ m

Factor of safety $= 2$

Ultimate strength $= 800$ kg

Weight of ice $= 910$ kg/m^3

The overall diameter of the conductor with ice coating is given by

$= d + 2t$

$= 19.5 + 2 \times 12.7$

$= 44.9$ mm

$= 4.49 \times 10^{-2}$ m

Wind load per meter length $=$ wind pressure $\times$ projected area

$= 4.49 \times 10^{-2} \times 39$

$= 1.751$ kg/m

Weight of ice per meter length $=$ Density of ice $\times \pi\, t(d+t)$

$= 910 \times \pi (12.7)(19.5 + 12.7)$

$= 910 \times \pi (408.94) \times 10^{-6}$

$= 910 \times 1284.72 \times 10^{-6}$

$= 1.169$ Kg/m

The resultant weight of the conductor, $W_r = \sqrt{(W_c + W_i)^2 + W_w^2}$

$$= \sqrt{(1+1.169)^2 + (1.751)^2}$$

$$= \frac{800}{2}$$

T   400 kgs

Permissible sag $S = \dfrac{W_r L^2}{8\,T}$

$$6.3 = \frac{2.78 \times L^2}{8 \times 400}$$

$$L^2 = \frac{6.3 \times 4 \times 400}{2.78}$$

$$L^2 = \frac{20160}{2.78}$$

$$L^2 = 7251.798$$

$$L = 85.16 \text{ mtrs.}$$

CHAPTER 7

Underground Cables

7.1 Introduction

Electrical power can be transmitted or distributed either by overhead system or by underground cable. Underground cables are used for transmission and distribution of power where it becomes impracticable to make use of overhead construction. Such location may be congested urban area where right-of-way cost would be excessive or local public prohibit overhead lines for reason of safety, or around plants and substations. The underground cables have several advantages such as less liable to damage through storms or lightning, low maintenance cost, less chance of faults, smaller voltage drop and good appearance. However, their major drawback is that they have greater installation cost and introduce insulation problems at high voltages compared with the equivalent overhead system.

The chief use of underground cables for many years has been for distribution of electric power in congested urban areas at comparatively low or moderate voltages. However, recent improvements in the design and manufacture have led to the development of cables suitable for use at high voltages. This has made it possible to employ underground cables for transmission of electric power for short and moderate distances.

An underground cable may be defined as one or more conductors covered with suitable insulation and surrounded by a protecting cover.

7.2 Properties of Insulating Material for Cable

In general, the insulating materials used in cables should have the following properties.

1. High insulation resistance to avoid leakage current.
2. High dielectric strength to avoid electrical breakdown of the cable.
3. High mechanical strength to withstand to the stresses.
4. Non-hygroscopic i.e., it should not absorb moisture from air or soil. The moisture tends to decreases the insulation resistance and hence the breakdown of the cable.
5. Non-inflammable.
6. Unaffected by acids and alkalies to avoid chemical reactions.
7. Low cost so as to make the underground system a viable proposition.
8. Low coefficient of thermal expansion.
9. Low permittivity
10. Capability of withstanding high rupturing voltages.

No one insulating material possess all the above mentioned properties, so the type of insulating material used in a cable depends upon the service for which the cable is required.

7.3 Requirements of an Underground Cable

Although several types of cables are available, the type of cable to be used will depend upon the working voltage and service requirements. In general, a cable must full fill the following requirements:

1. The conductor used in cables should be tinned stranded copper or aluminium of high conductivity. Stranding is done so that conductor may become flexible and carry more current.
2. The conductor size should be such that the cable carries the desired load current without overheating and causes voltage drop within permissible limits.
3. The cable must have proper thickness of insulation in order to give high degree of safety and reliability at the voltage for which it is designed.
4. The cable must be provided with suitable mechanical protection so that it may withstand the rough use in laying it.
5. The materials used in manufacture of cables should be such that there is complete chemical and physical stability throughout.

7.4 Classification of Cables

Cables for underground service may be classified in two ways according to

1. Type of insulating material used in their manufacture
2. The voltage for which they are manufactured.

Based on the operating voltage, cables can be divided into the following groups.

1. Low – tension (L.T.) cables – up to 1000 volts
2. High – tension (H.T.) cables – up to 11 KV
3. Super – tension (S.T.) cables – from 22 KV to 33 KV
4. Extra high – tension (E.H.T.) cables – from 33 KV to 66 KV
5. Extra Super voltage cables – beyond 132 KV

A cable may have one or more than one core depending upon the type of service for which it is intended. It may be single-core, two-core, three-core, four-core etc. For a 3-phase service, either 3-single-core cables or three-core cable can be used depending upon the operating voltage and load demand.

Cables for 3-phase service:

In practice, underground cables are generally required to deliver 3-phase power. For this purpose, either three-core cable or three single core cables may be used. For voltages up to 66 KV, 3-core cable is preferred due to economic reasons. However, for voltages beyond 66 KV, 3 core cables become too large therefore single core cables are used. The following cables are used for 3-phase service:

1. Belted Cables – up to 11 KV
2. Screened Cables – > 11 KV to < 66 KV
3. Pressure Cales – > 66 KV

7.5 Construction of a Single Core Cable

Single core cable has ordinary construction because the stresses developed in the cable for low voltages are small. It consists of one circular core of tinned stranded copper insulated by layers of impregnated paper. The insulation is surrounded by a lead sheath which prevents the entry moisture into the inner parts. In order to protect the

the lead sheath from corrosion, an overall serving of compounded fibrous material is provided. Single core cables are not usually armored in order to avoid excessive sheath losses. The principal advantage of single core cables is simple construction and availability of larger cross section.

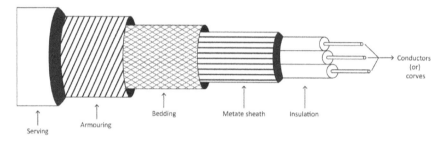

Fig. 7.1

(1) Cores (or) conductors:

Core is the central part of the cable which carries the actual current. A cable may have one or more than one cores depending on the type of service for which it is used. For instance, 3 conductor cables are shown in fig. which is used for 3ϕ service. The conductors are made of copper or aluminum and are usually stranded in order to increase the current carrying capacity and flexibility to the cable.

(2) Insulation:

Each core is provided with suitable thickness of insulation in order to prevent the leakage current. The thickness of insulation depends on the operating voltage of the cable. Higher the operating voltage, more will be the thickness of the insulation. The commonly used materials for insulation are rubber mineral compound. Impregnable paper, varnish cambric, polyvinyl chloride and XLPE.

(3) **Metallic sheath:**

A metallic sheath is provided around the insulation. The sheath is usually made up of lead or aluminum. The function of the metallic sheath is to protect the insulation from moisture, gases or other damaging liquids (acids or alkalies) in the soil. This can be protecting the insulation from direct contact with the soil. Moisture content and gases in the soil will deteriorate the insulation and also may cause some chemical reactions.

(4) **Bedding:**

Bedding is nothing but a layer of paper tape compounded with fibrous material provided over the metallic sheath. Paper tape compounded fibrous materials are Hessian tape (or) Jute. Bedding serves two purposes i.e., it protect the metallic sheath from corrosion due to moisture and it also protect from mechanical injury due to armouring.

(5) **Armouring:**

Armouring is provided over the bedding in order to provide good mechanical strength to the cable and to protect the cable from injuries during erection. It consists of layers of galvanized steel wire or steel tape. In order to reduce the sheath loss, the wire is made up of high resistance material.

(6) **Serving:**

In order to protect armouring from atmospheric conditions, a layer of fibrous material is provided over the armouring. This is known as "Serving".

7.6 Insulating Materials for Underground Cables

The satisfactory operation of a cable depends upon the characteristics of insulation used. Therefore, the proper choice of insulating material for cables is considerably important. The main requirements of the insulating materials used for cables are:

1. High insulation resistance to avoid leakage current.
2. High dielectric strength to avoid electrical breakdown of the cable.
3. Good mechanical properties i.e., tenacity and elasticity.
4. Immune to attacks by acids and alkalies over a range of temperature of about −18°C to 94°C
5. Non hygroscopic. i.e., it should not absorb moisture from air or soil. Moisture tends to decrease the insulation resistance. In case the insulating material is hygroscopic it must be enclosed in a water proof covering like lead sheath.
6. Non-inflammable.
7. Low coefficient of thermal expansion.
8. Capability to withstand high rupturing voltages.

No insulating material possesses all the above mentioned properties. Therefore, the type of insulating material used in a cable depends upon the service for which the cable is required. The various insulating materials used for cables are: Rubber vulcanized India rubber, Impregnated paper, varnished cambric, polyvinyl chloride and XLPE.

1. **Rubber:**

   Rubber may be obtained from milky sap of tropical trees or it may be produced from oil products. It has relative permittivity varying between 2 and 3, dielectric strength is about 30 KV/mm and resistivity of insulation is $10^{17} \Omega$ c although pure rubber has reasonably high insulating properties, it suffers from major drawbacks. Those are, it readily absorbs moisture, maximum safe temperature is low (about 38°C), soft and liable to damage due to rough handling and ages when exposed to light. Therefore, pure rubber cannot be used as an insulating material.

2. **Vulcanized India Rubber (VIR):**

   It is prepared by mixing pure rubber with mineral matter such as Sulphur, Zinc Oxide and red lead etc. the compound so formed is rolled into thin sheets and cut into strips. The rubber compound is then applied to the conductor and is heated to a temperature of about

150°C. The whole process is called Vulcanization and the product obtained is known as "VIR".

VIR has greater mechanical strength, durability and war resistant property than pure rubber. It's main drawback is that Sulphur reacts very quickly with copper and for this reason, cables using VIR insulation have tinned copper conductor. The VIR is used for low and moderate voltage cables.

3. **Impregnated paper:**

It consists of chemically pulped paper made from wood chippings and impregnated with some compound such as paraffinic or naphthenic material. It has the advantages of low cost, low capacitance, high dielectric strength and high insulation resistance. The only disadvantage is that paper is hygroscopic and even if it is impregnated with suitable compound, it absorbs moisture and thus lowers the insulation resistance of the cable. For this reason, paper insulated cables are always provided with some protective covering and are never left unsealed. It's dielectric strength is 30 KV/mm and insulation resistivity of the order 10^{-7} Ω.m.

Paper insulated cables are used for converging large blocks of power in transmission and distribution. It is particularly for distribution at low voltage in congested areas where thejoints are to be provided only at the terminal apparatus.

4. **Varnished Cambric:**

It s a cotton cloth impregnated and coated with varnish. This type of insulation is also known as "Empire tape". The cambric is lapped on to the conductor in the form of a tape and its surfaces are coated with petroleum jelly compound to allow for the sliding of one turn over another as the cable is bent. As the varnished cambric is hygroscopic, such cables are always provided with metallic sheath. It's dielectric strength is about 4 KV/mm and permittivity is 2.5 to 3.8.

5. **Polyvinyl Chloride (PVC):**

   This insulating material is a synthetic compound. It is obtained from the polymerisation of acetylene and is in the form of white powder. For obtaining this material as a cable insulation it is compounded with certain materials knows as plasticizers which are liquids with high boiling point. It is next to oxygen and many alkalies and acids. Therefore, its use is preferred over VIR in extreme environments such as in cement factory or chemical factory. It has high insulation resistance, good dielectric strength of 17 KV/mm dielectric constant (permittivity) of 5, good mechanical properties and maximum continuous temperature rating is 75°C. These are employee for low and medium voltage domestic and industrial light.

6. **XLPE (Cross-linked polythene):**

   It is low density polythene, when vulcanized under controlled conditions, results in cross-linking of carbon atoms and the compound produced is a new material having high melting point with light weight, small dimensions, low dielectric constant and good mechanical strength.

   XLPE cables have high maximum continuous temperature rating of 90° with dielectric strength of 20 KV/mm. Because of high temperature withstand capability and very low water absorption these cables are very suitable for all voltages up to 33 KV.

7.7 Insulation Resistance of a Cable

The opposition offered by the insulation to the leakage current is known as "Insulation resistance".

The cable conductor is provided with a suitable thickness of insulating material in order to prevent leakage current. The pat for leakage current is radial through the insulation. For the safety operation, the insulation resistance of the cable should be very high.

Consider a single core cable of conductor radius r_1 and internal sheath radius r_2. Length 'l' and insulation resistivity ρ.

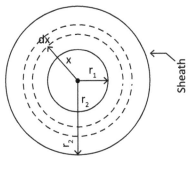

Fig. 7.2

Let an elementary cylindrical section of the insulation or radius 'x' and thickness 'dx' be considered. Now the length through which leakage current tends to flow is 'dx' and the area of cross-section offered to this current is $2\Pi x l$.

∴ Insulation resistance of the considered layer, $= \dfrac{\rho}{2\Pi x l} \cdot dx$

Insulation resistance of the whole cable is

$$R_{ins} = \int_{r_1}^{r_2} \dfrac{\rho}{2\Pi x l} \cdot dx = \dfrac{\rho}{2\Pi l} \int_{r_1}^{r_2} \dfrac{1}{x} \cdot dx$$

$$R_{ins} = \dfrac{\rho}{2\Pi l} \cdot \log_e \left(\dfrac{r_2}{r_1} \right)$$

This shows that insulation resistance of a cable is inversely propositional to its length. If the cable length increases its insulation resistance decreases and vice-versa.

7.8 Capacitance of a Single Core Cable

A single core cable can be considered to be equivalent to two long co-axial cylinders. The conductor or core of the cable is the inner cylinder while the outer cylinder is represented by lead sheath which is at earth potential.

Consider a single core cable with conductor diameter 'd' and inter sheath diameter 'D'. Let the charge per meter axial length of the cable 'Q' coulombs and 'ϵ' be the permittivity of the insulation material between core and lead sheath. $\epsilon = \epsilon_0 \epsilon_r$ where ϵ_r is the relative permittivity of the insulation.

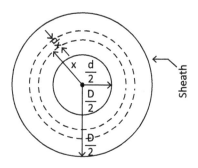

Fig. 7.3

Consider a cylinder of radius 'x' meters and axial length 1 meter. The surface area of this cylinder is $2\Pi x \times 1$ i.e. $2\Pi x$ m².

Field intensity at point 'P' is,

$$E_x = \frac{Q}{2\Pi \epsilon_0 \epsilon_r \cdot x} \text{ volts/m.}$$

The work done in moving a unit positive charge from point 'P' through a distance 'dx' in the direction of electric field is $E_x \, dx$. Hence, the work done in moving a unit positive charge from conductor to sheath, which is potential difference 'V' between conductor and sheath is,

$$V = \int_{d/2}^{D/2} E_x \cdot dx \Rightarrow \int_{\frac{d}{2}}^{\frac{D}{2}} \frac{Q}{2\Pi \epsilon_0 \epsilon_r \cdot x} \cdot dx$$

$$V = \frac{Q}{2\Pi \epsilon_0 \epsilon_r} \int_{d/2}^{D/2} \frac{1}{x} \, dx$$

$$V = \frac{Q}{2\Pi \epsilon_0 \epsilon_r} (\log_e x)_{d/2}^{D/2}$$

$$V = \frac{Q}{2\Pi \epsilon_0 \epsilon_r} \cdot \log_e \frac{D}{d}$$

Capacitance of a cable is

$$C = \frac{Q}{V} = \frac{Q}{\frac{Q}{2\Pi \epsilon_0 \epsilon_r} \cdot \log_e \left(\frac{D}{d}\right)}$$

$$C = \frac{2\Pi \epsilon_0 \epsilon_r}{\log_e \left(\frac{D}{d}\right)} \text{ F/m}$$

$$C = \frac{2\Pi \times 8.854 \times 10^{-12} \times \epsilon_r}{2.303 \log_{10}\left(\frac{D}{d}\right)} \text{ F/m}$$

$$C = \frac{\epsilon_r}{41.4 \log_{10}\left(\frac{D}{d}\right)} \times 10^{-9} \text{ F/m}$$

If the cable has a length 'l' meters, then capacitance of a cable is,

$$C = \frac{\epsilon_r \, l}{41.4 \log_{10}\left(\frac{D}{d}\right)} \times 10^{-9} \text{ F}$$

7.9 Dielectric Stress in a Single Core Cable

Under operating conditions, the insulation of a cable is subjected to electrostatic forces. This is known as "dielectric stress". The dielectric stress at any point in a cable is in fact the potential gradient (electrical field intensity) at that point.

Consider a single core cable with core diameter 'd' and internal sheath diameter 'D'. Electric fields intensity at a point 'x' meters from the centre of the cable is.

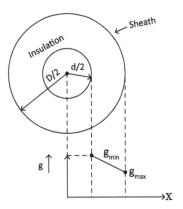

Fig. 7.4

$$E_x = \frac{Q}{2\Pi \epsilon_0 \epsilon_r \, x} \text{ volts/m}$$

By definition, electric field intensity is equal to potential gradient. Therefore, potential gradient 'g' at a point 'x' from the centre of cable is

$$g = E_x$$

$$\therefore g = \frac{Q}{2\Pi \epsilon_0 \epsilon_r \cdot x} \text{ V/m} \quad \rightarrow (1)$$

Now, potential difference between conductor and sheath is

$$V = \int_{d/2}^{D/2} E_x \, dx \Rightarrow \int_{d/2}^{D/2} \frac{Q}{2\Pi \epsilon_0 \epsilon_r \cdot x} \, dx$$

$$= \frac{Q}{2\Pi \epsilon_0 \epsilon_r} (\log_e x)_{d/2}^{D/2}$$

$$V = \frac{Q}{2\Pi \epsilon_0 \epsilon_r} \log_e \left(\frac{D}{d}\right) \text{ volts}$$

$$Q = \frac{2\Pi \epsilon_0 \epsilon_r \cdot V}{\log_e \left(\frac{D}{d}\right)} \rightarrow \quad \rightarrow (2)$$

Substitute equation (2) in equation (1), then

$$g = \frac{1}{2\Pi \epsilon_0 \epsilon_r \cdot x} \cdot \frac{2\Pi \epsilon_0 \epsilon_r \cdot V}{\log_e \left(\frac{D}{d}\right)}$$

$$g = \frac{V}{x \cdot \log_e \left(\frac{D}{d}\right)} \text{ V/m} \quad \rightarrow (3)$$

It is clear from equation (3), that the potential gradient varies inversely as distance 'x'. Therefore, potential gradient will be maximum when 'x' is minimum i.e., $x = \frac{d}{2}$ or at the surface of the conductor on the other hand, potential gradient will be minimum at $x = \frac{D}{2}$ or at sheath surface.

∴ Maximum potential gradient is,

$$g_{max} = \frac{V}{\frac{d}{2} \cdot \log_e \left(\frac{D}{d}\right)} \quad \left[\because x = \frac{d}{2}\right]$$

$$g_{max} = \frac{2V}{d \cdot \log_e \left(\frac{D}{d}\right)} \text{ volts/m}$$

$$g_{min} = \frac{V}{\frac{D}{2} \cdot \log_e\left(\frac{D}{d}\right)}$$

$$g_{min} = \frac{2V}{D \cdot \log_e\left(\frac{D}{d}\right)} \text{ volts/m}$$

7.10 Most Economical Size (Or Diameter) of Conductor

For a given voltage and overall diameter of a single core cable, there is a certain value of the conductor radius or diameter that gives a minimum potential gradient at the surface. A small conductor radius allows a greater insulator thickness but on the other hand, the smaller radius of curvature tends to increase the stress.

For fixed values of voltage 'V' and overall diameter 'D' the g_max will be minimum when $\left[d\log_e\left(\frac{D}{d}\right)\right]$ is maximum and $\left[d\log_e\left(\frac{D}{d}\right)\right]$ will be maximum if

$$\frac{d}{dd}\left[d\log_e\left(\frac{D}{d}\right)\right] = 0$$

$$d\frac{d}{dd}\log_e\left(\frac{D}{d}\right) + \log_e\left(\frac{D}{d}\right)\frac{d}{dd}(d) = 0$$

$$d.\left(\frac{d}{D}\right)\frac{d}{dd}\left(\frac{D}{d}\right) + \log_e\left(\frac{D}{d}\right)(1) = 0$$

$$\left(\frac{d^2}{D}\right) \cdot \frac{d(0) - D(1)}{d^2} + \log_e\left(\frac{D}{d}\right) = 0$$

$$\left(\frac{d^2}{D}\right) \cdot \left(-\frac{D}{d^2}\right) + \log_e\left(\frac{D}{d}\right) = 0$$

$$-1 + \log_e\left(\frac{D}{d}\right) = 0 \Rightarrow \log_e\left(\frac{D}{d}\right) = 1$$

$$\frac{D}{d} = e^1 \Rightarrow d = \frac{D}{e^1} = \frac{D}{2.71828} \text{ and}$$

$$g_{max} = \frac{2V}{d} \text{ V/m}$$

For the low and medium voltage cables, the diameter of the conductor obtained from the above consideration is,

$$d = \frac{2V}{g_{max}}$$

$d = \dfrac{2V}{g_{max}}$ is some what less than that obtained from the consideration of the safe current density. Therefore, the main criterion for determining conductor diameter for such cables is the current carrying capacity.

Problems

1. **Find the most economical value of the diameter of a single core cable to be used on a 132 KV, 3ϕ system. Also find overall diameter of the insulation if the peak permissible stress is not to exceed 60 KV/cm.**

    Maximum stress, $g_{max} = 60$ KV/cm (peak)

    RMS value of phase voltage, $V = \dfrac{132}{\sqrt{3}} = 76.21$ KV

    Peak value of phase voltage, $V = \sqrt{2} \times 76.21$

    $= 107.76$ KV

    Most economical diameter of conductor, $d = \dfrac{2V}{g_{max}}$

    $$d = \frac{2 \times 107.76 \times 10^3 \times 10^{-2}}{60 \times 10^3}$$

    $d = 0.0359$ meters

    Internal sheath diameter $D = d \times 2.711828 \quad \left[\because d = \dfrac{D}{2.718} \right]$

    $= 0.0359 \times 2.71828$

    $= 0.09764$ metr

2. **A single core cable has a conductor diameter of 1 cm and insulation thickness of 0.4 cm. if the specific resistance of insulation is 5×10^{14} Ʊ.cm . Calculate the insulation resistance for a 2 Km length of the cable.**

Conductor Radius, $r_1 = \dfrac{1}{2} = 0.5$ cm

Length of the cable, $l = 2$ km $= 2000$ m

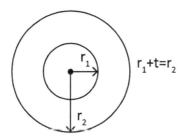

Insulation Resistivity, $\rho = 5 \times 10^{14} \Omega\text{-cm} = 5 \times 10^{12} \Omega\text{-cm}$

Internal sheath radius, $r_2 = r_1 + t = 0.5 + 0.4 = 0.9$ cm

∴ Insulation resistance of cable is,

$$R_{ins} = \dfrac{1}{2\pi l} \log_e \left(\dfrac{r_2}{r_1}\right)$$

$$= \dfrac{5 \times 10^2}{2\pi \times 2000} \log_e \left(\dfrac{0.9}{0.5}\right)$$

$R_{ins} = 2.34$ MΩ

7.11 Grading of Cables

The process of achieving uniform electrostatic stress in the dielectric of cables is known as "grading of cables".

Electrostatic stress in a single core cable has a maximum value at the conductor surface and goes on decreasing as we move towards the sheath. The maximum voltage that can be safely applied to a cable depends upon g_{max} i.e., electrostatic stress at the conductor surface. For safe working of a cable having homogeneous dielectric, the strength of dielectric must be more than g_{max}. If a dielectric of high strength is used for a cable, it is useful only near the conductor where stress is maximum. But as we move away from the conductor, the electrostatic stress decreases, so dielectric will be unnecessarily over strong. This leads to unequal stress distribution in a cable.

The unequal stress distribution in a cable is undesirable for two reasons.

1. Insulation of greater thickness is required which increases the cable size.
2. It may lead to the breakdown of insulation.

In order to overcome above disadvantages, it is necessary to have a uniform stress distribution in cables. This can be achieved by distributing the stress in such a way that its value is increased in the outer layers of dielectric. This is known as "grading of cables". The following are two main methods of grading of cables:

1. Capacitance grading
2. Inter sheath grading

7.11.1 Capacitance Grading

The process of achieving uniformly in the dielectric stress by using layers of different dielectrics is known as "capacitance grading".

In capacitance grading, the homogeneous dielectric is replaced by a composite dielectric. The composite dielectric consists of various layers of different dielectrics in such a manner that relative permittivity ϵ_r of any layer is inversely propositional to its distance from the centre. Under such conditions, the value of potential gradient at any point in the dielectric is constant and is independent of its distance from the centre. In other words, the dielectric stress is same everywhere and the grading is ideal one. However, ideal grading requires the use of infinite number of dielectrics which is an impossible task. In practice, two or three dielectrics are used in the decreasing order of permittivity. The dielectric of highest permittivity being used near the core.

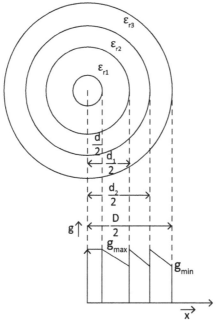

Fig. 7.5

The capacitance grading can be explained by above fig. There are three dielectrics of outer diameter d_1, d_2 and D and of relative permittivity ϵ_1, ϵ_2 and ϵ_3 respectively. If the primitivities are such that $\epsilon_{r_1} > \epsilon_{r_2} > \epsilon_{r_3}$ and three dielectrics are worked at the same maximum stress, then

$$\frac{1}{\epsilon_{r_1} d} = \frac{1}{\epsilon_{r_2} d_1} = \frac{1}{\epsilon_{r_3} d_2}$$

$$\epsilon_{r_1} d = \epsilon_{r_2} d_1 = \epsilon_{r_3} d_2$$

Potential difference across the inner layer is,

$$V_1 = \int_{d/2}^{d_1/2} g\, dx \qquad [\because E = g]$$

$$= \int_{d/2}^{d_1/2} \frac{Q}{2\Pi\, \varepsilon_0 \varepsilon_{r_1} \cdot x}\, dx$$

$$= \frac{Q}{2\Pi\, \varepsilon_0 \varepsilon_{r_1}} \int_{d/2}^{d_1/2} \frac{1}{x}\, dx$$

$$V_1 = \frac{Q}{2\pi\varepsilon_0\varepsilon_{r_1}}\log_e\left(\frac{d_1}{d}\right)$$

$$V_1 = \frac{g_{max}}{2} \cdot d\log_e\left(\frac{d_1}{d}\right)$$

Similarly, Potential difference across second layer (V_2) and third layer (V_3) is given by,

$$V_2 = \frac{g_{max}}{2} \cdot d_1 \log_e\left(\frac{d_2}{d_1}\right)$$

$$V_3 = \frac{g_{max}}{2} \cdot d\log_e\left(\frac{D}{d_2}\right)$$

Total potential difference between core and sheath is

$$V = V_1 + V_2 + V_3$$

$$V = \frac{g_{max}}{2}\left[d\log_e\left(\frac{d_1}{d}\right) + d_1 \log_e\left(\frac{d_2}{d_1}\right) + d_2 \log_e\left(\frac{D}{d_2}\right)\right]$$

$$g = \frac{Q}{2\Pi\epsilon_0\epsilon_r \cdot x}$$

$$g_{max} = \frac{Q}{2\Pi\epsilon_0\epsilon_r \cdot \left(\frac{d}{2}\right)}$$

$$g_{max} = \frac{Q}{\Pi\epsilon_0\epsilon_r \cdot d}$$

$$g_{max} \cdot d = \frac{Q}{\Pi\epsilon_0\epsilon_r}$$

$$\frac{g_{max} \cdot d}{2} = \frac{Q}{2\Pi\epsilon_0\epsilon_r}$$

If the cable had homogeneous dielectric, then for the same values of d, D and g_{max}, the permissible potential difference between core and sheath would have been

$$V^l = \frac{g_{max}}{2} \cdot d\log_e\left(\frac{D}{d}\right)$$

Obviously $V > V^l$ i.e., for given dimensions of the cable, a graded cable can be worked at a greater potential than non-graded cable. For the same safe potential, the size of graded cable will be less than that of non graded cable.

The following points may be noted:

i. As the permissible values of g_{max} are peak values, all the voltages in above expressions should be taken as peak values and not the rms values.

ii. If the max. Stress in the three dielectrics is not the same, then

$$V = \frac{g_{1max}}{2} \cdot d\log_e\left(\frac{d_1}{d}\right) + \frac{g_{2max}}{2} \cdot d_1 \log_e\left(\frac{d_2}{d_1}\right) + \frac{g_{3max}}{2} \cdot d_2 \log_e\left(\frac{D}{d_2}\right)$$

7.11.2 Inter Sheath Grading

In this method of cable grading, a homogeneous dielectric is used, but it is divided into various layers by placing metallic inter sheath between the core and lead sheath. The inter sheaths are held at suitable potentials which are in between the core and earth potential. This arrangement improves voltage distribution in the dielectric of the cable and consequently more uniform potential gradient is obtained.

Consider a cable of core diameter 'd' and outer lead sheath of diameter D. suppose that two inter sheaths of diameter d_1 and d_2 are inserted into the homogeneous dielectric and maintained at some fixed potentials. Let V_1, V_2 and V_3 respectively be the voltage between core and inter sheath 1, between inter sheath 1 and 2 and between inter sheath 2 and outer load sheath. As these is a definite potential difference between the inner and outer layers of each inter sheath, each inter sheath can be treated like a homogenous single core cable. The above arrangement is shown in below fig.

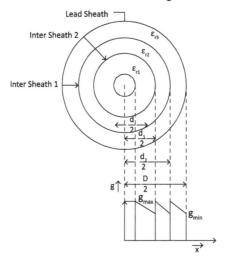

Fig. 7.6

Maximum stress between core and inter sheath 1 is,

$$g_{1max} = \frac{V_1}{\frac{d}{2} \cdot \log_e\left(\frac{d_1}{d}\right)}$$

Similarly, $g_{2max} = \dfrac{V_2}{\dfrac{d_1}{2} \cdot \log_e\left(\dfrac{d_2}{d_1}\right)}$

$$g_{3max} = \frac{V_3}{\frac{d_2}{2} \cdot \log_e\left(\frac{D}{d_2}\right)}$$

Since the dielectric is homogeneous, the maximum stress in each layer is the same i.e.,

$$g_{1max} = g_{2max} = g_{3max} = g_{max} \text{ (say)}$$

$$\therefore \frac{V_1}{\frac{d}{2} \cdot \log_e\left(\frac{d_1}{d}\right)} = \frac{V_2}{\frac{d_1}{2} \cdot \log_e\left(\frac{d_2}{d_1}\right)} = \frac{V_3}{\frac{d_2}{2} \cdot \log_e\left(\frac{D}{d_2}\right)}$$

As the cable behaves like three capacitors in series, therefore all the potentials are phase i.e., voltage between conductor and sheath is,

$$V = V_1 + V_2 + V_3$$

The inter sheath grading has the following disadvantages:

1. There is a possibility of damage to inter sheath during laying of the cable since the inter sheath has to be thin.
2. There are considerable losses in the inter sheaths due to charging current.
3. There are complications in fixing the sheath potentials.

Problems

3. A single core lead sheathed cable is graded by using 3 dielectrics of relative permittivity 5, 4 and 3 respectively. The conductor diameter is 2 cm and overall diameter is 8 cm. if 3 dielectrics are worked at the

same max. stress of 40 KV/cm. Find the safe working voltage for an ungraded cable.

What will be the safe working voltage for an ungraded cable, assuming the same conductor and overall diameter 3 max. stress?

Sol: Conductor diameter $d = 2$ cm

Sheath diameter $D = 8$ cm

Diameter of 1st dielectric $d_1 = ?$ and

Diameter of 2nd dielectric $d_2 = ?$

Relative permittivity's of dielectric are $\epsilon_{r_1} = 5$, $\epsilon_{r_2} = 4$, $\epsilon_{r_3} = 3$

Max. dielectric stress $g_{max} = 40 \, KV/cm$

Graded cable:

As the maximum stress in the 3 dielectrics is same,

$\epsilon_{r1} d = \epsilon_{r2} d_1 = \epsilon_{r3} d_2$

$5 \times 2 = 4 \times d_1 = 3 \times d_2$

$10 = 4 d_1 \Rightarrow d_1 = 2.5 \, cm$

$10 = 3 d_2 \Rightarrow d_2 = 3.3 \, cm$

Permissible peak voltage for the cable

$$= \frac{g_{max}}{2} \left[d \log_e \left(\frac{d_1}{d}\right) + d_1 \log_e \left(\frac{d_2}{d_1}\right) + d_2 \log_e \left(\frac{D}{d_2}\right) \right]$$

$$= \frac{40}{2} \left[2 \log_e \left(\frac{2.5}{2}\right) + 2.25 \log_e \left(\frac{3.34}{2.25}\right) + 3.34 \log_e \left(\frac{8}{3.34}\right) \right]$$

$= 20 \times 4.0904$

$= 81.808 \, KV$ (Peak)

Safe working voltage (rms) for cable $= \dfrac{81.808}{\sqrt{2}} = 57.84 \, KV$

Ungraded cable:

Permissible peak voltage for the cable $= \dfrac{g_{max}}{2} \cdot d \log_e \left(\dfrac{D}{d}\right)$

$= \dfrac{40}{2} \cdot 2 \log_e \dfrac{8}{2}$

$= 55.4 \, KV$

Safe r.m.s working voltage $= \dfrac{55.4}{\sqrt{2}}$

$= 39.2 \, KV$

4. A single core cable of conductor diameter 2 cm and lead sheath of diameter 5.3 cm is to be used on a 6 KV, 3ϕ system. Two inter sheaths of diameter 3.1 cm and 4.2 cm are introduced between the core and sheath. If the max. stress in the layers is the same, find the voltages on the inter sheaths.

Sol: Conductor diameter $d = 2 \, cm$

Diameter of 1st inter sheath $d_1 = 3.1 \, cm$

Diameter of 2nd inter sheath $d_2 = 4.2 \, cm$

Diameter of lead sheath $D = 5.3 \, cm$

Operating voltage $V = 66 \, KV(r.m.s) \, 3\phi$ system

Operating voltage for single core cable, $V = \dfrac{66}{\sqrt{3}} \, (r.m.s)$

Operating voltage in peak for single core cable $= \dfrac{66}{\sqrt{3}} \times \sqrt{2}$ (peak)

$= 53.9 \, KV$

Maximum stress between core and inter sheath 1 is,

$g_{1max} = \dfrac{V_1}{\dfrac{d}{2} \cdot \log_e \left(\dfrac{d_1}{d}\right)}$

$= \dfrac{V_1}{1 \times \log_e \left(\dfrac{3.1}{2}\right)}$

$g_{1max} = 2.28 \, V_1$

Max. stress between inter sheath 1 & inter sheath 2 is,

$$g_{2max} = \frac{V_2}{\frac{d_1}{2} \cdot \log_e\left(\frac{d_2}{d_1}\right)}$$

$$g_{2max} = \frac{V_2}{1.55 \times \log_e\left(\frac{4.2}{3.1}\right)} = 2.12\, V_2$$

Max. stress between inter sheath 2 and lead sheath

$$g_{3max} = \frac{V_3}{\frac{d_2}{2} \cdot \log_e\left(\frac{D}{d_2}\right)}$$

$$= \frac{V_3}{2.1 \log_e\left(\frac{5.3}{4.2}\right)}$$

$$g_{3max} = 2.04\, V_3$$

As the max. stress in the layers is same

$$g_{1max} = g_{2max} = g_{3max}$$

$$2.28\, V_1 = 2.12\, V_2 = 2.04\, V_3$$

$$V_2 = \frac{2.28}{2.12} V_1$$

$$V_2 = 1.075\, V_1$$

$$V_3 = \frac{2.28}{2.04} V_1$$

$$V_3 = 1.117\, V_1$$

Now $V = V_1 + V_2 + V_3$

$$= V_1 + 1.075\, V_1 + 1.117\, V_1$$

$$= 53.9 = 3.192\, V_1$$

$$V_1 = \frac{53.9}{3.192}$$

$V_1 = 16.88\ KV, \quad V_2 = 1.075\ V_1 = 1.075 \times 16.88$

$V_2 = 18.14\ KV$

∴ Voltage on 1ˢᵗ inter sheath $= V - V_1$
$$= 53.9 - 16.88 = 37.02\ KV$$

Voltage on 2ⁿᵈ inter sheath $= V - V_1 - V_2 = 53.9 - 16.8 - 18.14$
$$= 18.88\ KV$$

5. **The insulation resistance of a single core cable is $495\ M\Omega/m$. If the core diameter is $2.5\ cm$ and resistivity of insulation is $4.5 \times 10^{24}\ \Omega.cm$. Find the insulation thickness.**

Sol: Length of cable $l = 1\ Km = 1000\ m$

Insulation resistance of cable, $R = 495\ M\Omega/m$

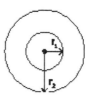

Conductor radius $r_1 = \dfrac{2.5}{2} = 1.25\ cm$

Resistivity of insulation $\rho = 4.5 \times 10^{14}\ \Omega-cm \Rightarrow 4.5 \times 10^{12}\ \Omega-m$.

Let r_2 be the sheath radius,

Insulation resistance $R = \dfrac{\rho}{2\Pi l} \log_e \left(\dfrac{r_2}{r_1}\right)$

$$495 \times 10^6 = \dfrac{4.5 \times 10^{12}}{2\Pi \times 1} \times \log_e \left(\dfrac{r_2}{r_1}\right)$$

$$\log_e\left(\frac{r_2}{r_1}\right) = 0.69$$

$$\frac{r_2}{r_1} = e^{(0.69)}$$

$$\frac{r_2}{r_1} = 1.9937$$

$$r_2 = 1.9937 \times 1.25$$

$$r_2 \cong 2.5 \text{ cm}$$

Insulation thickness $t = r_2 - r_1$

$$= 2.5 - 1.25$$

$$= 1.25 \text{ cm}$$

7.12 Capacitance of 3-Core Belted Cables

The capacitance of a cable system is of much greater importance than that of overhead line of the same length first due to the narrow spacing between the conductors themselves and between the conductors and earthed sheath. Secondly due to their separation by a dielectric medium of higher permittivity as compared to air.

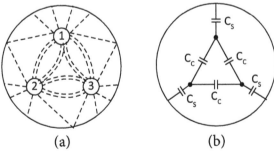

Fig. 7.7

Since there is a potential difference exists between pairs of conductors and between each conductor and sheath, electrostatic fields are set up in the cable as shown in fig (1). These electrostatic fields give rise to core-core capacitance 'C_c' and conductor – earth capacitance (or) conductor to sheath capacitance 'C_s' as shown in fig (2).

They lay of a belted cable makes it reasonable to assume equality of each C_c and each C_s.

The 3 delta connected capacitance C_c can be converted into equivalent star connected capacitances as shown in fig (3).

NOTE: The equivalent star capacitance C_{eq} is equal to 3 times the delta capacitance C_c. i.e., $C_{eq} = 3\ C_c$.

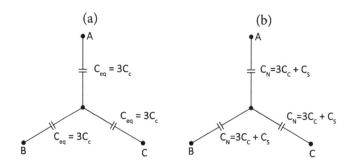

Fig. 7.8

The star point may be assumed to be at zero potential and if the sheath is also at zero potential, the capacitance of each core to neutral is,

$$C_N = C_{eq} + C_S$$

$$C_N = 3\ C_C + C_S$$

If V_P is the phase voltage, then the charging current I_C is given by,

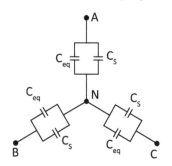

Fig. 7.9

$$I_C = 2\ \Pi\ f C_N V_{Ph}$$

$$I_C = 2\ \Pi\ f\left(3\ C_c + C_s\right) V_{Ph}\ \text{Ampers}$$

The value of capacitances C_c and C_s may be obtained by measurement as follows.

1. Capacitance is measured between the three cores bunched together and earthed sheath. This gives $3\,C_s$ because three capacitances C_c are eliminated leaving the three capacitances C_s in parallel. Therefore, if C_1 is measured capacitance, then

$$C_1 = 3\,C_s$$

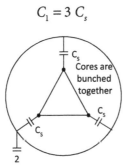

Fig. 7.10

$$\text{or,}\ C_s = \frac{C_1}{3}$$

2. Capacitance between the two cores or lines is measured with the third core being either insulated or connected to sheath as shown fig. this eliminates one of the capacitors 'C_s'. If 'C_2' is measured capacitance, then

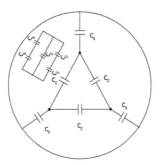

Fig. 7.11

$$C_2 = C_C + \frac{C_C}{2} + \frac{C_S}{2}$$

$$= \frac{1}{2}(2C_C + C_C + C_S)$$

$$C_2 = \frac{1}{2}(3C_C + C_S)$$

$$= \frac{1}{2} \cdot C_N$$

$$C_2 = \frac{1}{2} C_N$$

$$C_2 = C_C + \frac{C_S \times C_S}{C_S + C_S} + \frac{C_C \times C_C}{C_C + C_C}$$

$$C_2 = C_C + \frac{C_S}{2} + \frac{C_C}{2}$$

3. Two cores are bunched with sheath and capacitance is measured between them and the third core as shown in fig. if measured capacitance is C_3, then

$$C_3 = 2C_C + C_S$$

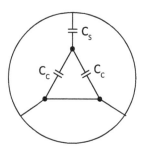

Fig. 7.12

From the above equations, the values of C_C and C_S can be obtained

6. **A single-core cable has a conductor diameter of $1\,cm$ and insulation thickness of $0.4\,cm$. If the specific resistance of insulation is $5 \times 10^{-4}\,\Omega.cm$. Calculate the insulation resistance for $2\,km$ length of the cable?**

A. Given data:

Diameter of the core, $d = 1\ cm$

Insulation thickness, $t = 0.4\ cm$

Specific resistance of insulation, $\rho = 5 \times 10^{14}\,\Omega.m$

Length of the cable, $l = 2$ cm

Insulating resistance, $R_{ins} = \dfrac{\rho}{2\Pi l}\left[\ln\left(\dfrac{D}{d}\right)\right]$

Diameter of the cable, $D = d + 2t$

$$= 1 + 2 \times 0.4$$
$$= 1.8 \text{ cm}$$

$$R_{ins} = \dfrac{5 \times 10^{14}}{2\Pi \times 2 \times 10^3 \times 10^2}\ln\left(\dfrac{1.8}{1}\right)$$

$$R_{ins} = 233.87 \ M\Omega$$

7. **Find the most economical value of the diameter of single. Core cable to be used on $132\,KV$, $3-\phi$ system also find over all diameter of the internal sheath if the peak permissible stress is not to exceed $60\,KV/cm$.**

A. Given data:

Line voltage, $V_L = 132$ V

Phase voltage $V_{Ph} = \dfrac{V_L}{\sqrt{3}} = \dfrac{132}{\sqrt{3}} = 76.2$ KV

Maximum stress, $g_{\max(peak)} = 60$ KV/cm

$$(g_{\max})(RMS) = \sqrt{2} \times \dfrac{60}{\sqrt{3}} = 42.426 \ KV/cm$$

∴ Maximum gradient, $g_{\max} = \dfrac{2V}{d}$

$$d = \dfrac{2V}{g_{\max}} = \dfrac{2 \times 76.21}{42.426} = 3.59 \text{ cm}$$

The core diameter, $d = \dfrac{D}{2.778}$

∴ Overall diameter of the sheath, $D = d \times 2.778$

$$D = 3.59 \times 2.778$$
$$D = 9.75 \text{ cm}$$

8. A single-core lead sheathed cable is graded by dielectrics of relative permittivity 5, 4 & 3 respectively. the conductor diameter is $2\,cm$ and overall diameter is $8\,cm$. If the 3 dielectrics are works at the same maximum stress of $40\,KV/cm$. Find the safe working voltage of the cable? What will be the safe working voltage for an ungraded cable assuming the same conductor over all diameter, and maximum stress?

A. Given data:

Conductor diameter, $d = 2$ cm

Overall diameter, $D = 8$ cm

The relative permittivity of the 1st dielectric, $\epsilon_{r1} = 5$

The relative permittivity of the 2nd dielectric, $\epsilon_{r2} = 4$

The relative permittivity of the 3rd dielectric, $\epsilon_{r3} = 3$

As the maximum stress in 3 dielectrics is same, so

$\epsilon_{r1}d = \epsilon_{r2}d_1 = \epsilon_{r3}d_2$

$\epsilon_{r1}d = \epsilon_{r2}d \qquad \epsilon_{r1}d = \epsilon_{r3}d_2$

$5 \times 2 = 4 \times d_1 \qquad 5 \times 2 = 3 \times d_2$

$d_1 = 2.5$ cm $\qquad d_2 = 3.33$ cm

For graded cable:

Safe working voltage, $V = \dfrac{g_{max}}{2}\left[d\ln\left(\dfrac{d_1}{d}\right) + d_1\ln\left(\dfrac{d_2}{d_1}\right) + d_2\ln\left(\dfrac{D}{d_2}\right)\right]$

$= \dfrac{40}{2}\left[2\ln\left(\dfrac{2.5}{2}\right) + 2.5\ln\left(\dfrac{3.33}{2.5}\right) + 3.33\ln\left(\dfrac{8}{3.33}\right)\right]$

$V = 81.63\,KV\,(peak)$

In RMS, $V = \dfrac{81.63}{\sqrt{2}} = 57.72\,KV\,(RMS)$

For graded cable;

Safe working voltage, $V = \dfrac{g_{max}}{2} d \ln\left(\dfrac{D}{d}\right)$

$= \dfrac{40}{2} \times 2 \times \ln\left(\dfrac{8}{2}\right)$

$V = 55.45 \; KV \; (Peak)$

In RMS, $V = \dfrac{55.45}{\sqrt{2}} = 39.21 \; KV \; (RMS)$

9. **A single - core cable of conductor dia 2 cm and lead sheath dia 5.3 cm is to be used on 0.66 KV, 3-ϕ system. 2 inter sheaths of dia 3.1cm & 4.2 cm are introduced between the core and sheath if the maximum stress in the layers is the same find the voltages on the inter sheaths?**

A. Given data:

Diameter of the conductor, $d = 2cm$

Diameter of the lead sheath, $D = 5.3 \; cm$

Line voltage, $V_L = 66 \; KV \; (rms)$

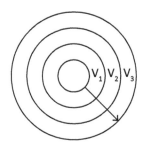

$V_{Ph} = \dfrac{V_L}{\sqrt{3}} = \dfrac{66}{\sqrt{3}} = 38.105 \; KV$

$V_{Ph_{peak}} = \dfrac{38.105}{\sqrt{2}} \times \sqrt{2} = 53.88 \; KV$

Diameter of inter sheath – 1, $d_1 = 3.1 \; cm$

Diameter of inter sheath – 2, $d_2 = 4.2\ cm$

Maximum stress, $g_{max} = \dfrac{V}{\dfrac{d}{2}\ln\left(\dfrac{D}{d}\right)}$

Because of inter sheath grading maximum stress is remain same at all sheaths.

$\therefore\ g_{1max} = g_{2max} = g_{3max}$

$\therefore$ Maximum stress of sheath – 1, $g_{1max} = \dfrac{V_1}{\dfrac{d}{2}\ln\left(\dfrac{d_1}{d}\right)}$

$$g_{1max} = \dfrac{V_1}{\dfrac{2}{2}\ln\left(\dfrac{3.1}{2}\right)}$$

$g_{1max} = 2.28\ V_1$

Maximum stress of sheath – 2, $g_{2max} = \dfrac{V_2}{\dfrac{d_1}{2}\ln\left(\dfrac{d_2}{d_1}\right)}$

$$= \dfrac{V_2}{\dfrac{3.1}{2}\ln\left(\dfrac{4.2}{3.1}\right)}$$

$g_{2max} = 2.12\ V_2$

Maximum stress of sheath – 3, $g_{3max} = \dfrac{V_3}{\dfrac{d_2}{2}\ln\left(\dfrac{D}{d_2}\right)}$

$$= \dfrac{V_3}{\dfrac{4.2}{2}\ln\left(\dfrac{5.3}{4.2}\right)}$$

$g_{3max} = 3.04\ V_3$

$\therefore\ 2.28\ V_1 = 2.12\ V_2 = 2.04\ V_3$

$$2.28\ V_1 = 2.12\ V_2 \qquad 2.28\ V_1 = 2.04\ V_3$$

$$V_2 = 1.075\ V_1 \qquad V_3 = 1.11\ V_1$$

Now, total voltage $V = V_1 + V_2 + V_3 = 53.88\ KV$

$$53.88 = V_1 + 1.075\ V_1 + 1.11\ V_1$$

$$53.88 = 3.185\ V_1$$

$$V_1 = 16.91\ KV$$

$$V_2 = 16.91 \times 1.075 = 18.17\ KV$$

$$V_3 = 1.11 \times 16.91 = 18.77\ KV$$

Voltage on 1st sheath, $V_1 = V - V_1$

$$= 53.88 - 16.91$$

$$V_1 = 36.97\ KV$$

Voltage on 2nd inter sheath, $V_2 = V - (V_1 + V_2)$

$$= 53.88 - (16.91 + 18.17)$$

$$V_2 = 18.8\ KV$$

10. The insulation resistance of a single core cable is $495\ M\Omega/m$. If the core diameter is $2.5\ cm$ and resistivity of insulation is $4.5 \times 10^{14}\ \Omega.cm$. Find the insulation thickness?

A. Given data:

Insulation resistivity, $\rho = 4.5 \times 10^{14}\ \Omega.cm$

$$= 4.5 \times 10^{12}\ \Omega m$$

Diameter of the core, $d = 2.5\ cm$, $r = \dfrac{2.5}{2} = 1.25\ cm$

Insulation resistance, $R_{ins} = 495\ M\Omega/m$

Assume length of cable, $l = 1\ km = 1000\ m$

Insulation resistance, $R_{ins} = \dfrac{\rho}{2\Pi l}\left[\ln\left(\dfrac{r_2}{r_1}\right)\right]$

$$495\times 10^6 = \dfrac{4.5\times 10^{12}}{2\Pi \times 1000}\ln\left(\dfrac{r_2}{2.5}\right)$$

$$\ln\left(\dfrac{r_2}{2.5}\right) = 6.91\times 10^{-4}$$

$$\dfrac{r_2}{2.5} = e^{6.91\times 10^{-4}}$$

$$r_2 = 1.00\times 2.5$$

$$r_2 = 2.5\ cm$$

∴ The thickness of the insulation resistance, $t = r_2 - r_1$

$$t = 2.5 - 1.25$$

$$t = 1.25\ cm$$

11. **A 11KV, single core cable has a conductor diameter of 20mm and sheath of inside diameter of 80mm. Find the maximum & minimum stress in the insulation and also find ratio of maximum to minimum stress?**

A. Given data:

Conductor diameter, $d = 20\ mm$

Sheath diameter, $D = 80\ mm$

Line voltage, $V_c = 11\ KV$

$$V_{Ph} = \dfrac{V_L}{\sqrt{3}} = 6.35\ KV$$

$$V_{Ph(peak)} = 6.35\times \sqrt{2} = 8.98\ KV$$

Maximum stress, $g_{max} = \dfrac{V}{\dfrac{d}{2}\ln\left(\dfrac{D}{d}\right)} = \dfrac{8.98}{\dfrac{20}{2}\ln\left(\dfrac{80}{20}\right)}$

$$g_{max} = 0.647\ KV/mm$$

Minimum stress, $g_{min} = \dfrac{2V}{D \ln\left(\dfrac{D}{d}\right)} = \dfrac{2 \times 8.98}{80 \ln\left(\dfrac{80}{20}\right)} = 0.169 \ KV/mm$

$g_{min} = 0.169 \ KV/mm$

Ratio of max to min stress $= \dfrac{g_{max}}{g_{min}} = \dfrac{\dfrac{2V}{d \ln\left(\dfrac{D}{d}\right)}}{\dfrac{2V}{D \ln\left(\dfrac{D}{d}\right)}} = \dfrac{D}{d}$

$= \dfrac{80}{20} = 4$

12. The inner and outer diameter of a cable are $3\,cm$ & $8\,cm$. The cable is insulated with 2 materials having permittivity of 5 & 3.5 with corresponding stress of $38\,KV/cm$ and $30\,KV/cm$. Calculate the radial thickness of each insulating layer & the safe working voltage of the cable?

A. Given data:

Inner diameter of the cable, $d = 3 \ cm$, $r = 1.5 \ cm$

Outer diameter of cable, $D = 8 \ cm$, $R = 4 \ cm$

Relative permittivity of 1st material, $\epsilon_{r_1} = 5$

Relative permittivity of 2nd material, $\epsilon_{r_2} = 3.5$

Maximum stress for material, $g_{max1} = 38 \ KV/cm$

Maximum stress for material, $g_{max2} = 30 \ KV/cm$

$\dfrac{g_{1\,max}}{g_{2\,max}} = \dfrac{\dfrac{q}{2\Pi \epsilon_1 r}}{\dfrac{q}{2\Pi \epsilon_2 r_2}}$

$$\frac{g_1}{g_2} = \frac{\epsilon_2 r_1}{\epsilon_1 r}$$

$$\frac{38}{30} = \frac{3.5 \times r_1}{5 \times 1.5}$$

∴ Radius of inner dielectric, $r_1 = 2.7$ cm

Radial thickness of inner dielectric $= 2.7 - 1.5 = 1.2$ cm

Radial thickness of outer dielectric $= 4 - 2.7 = 1.3$ cm

Maximum voltage of the cable $= g_{1\,max} \cdot r.\ln\dfrac{r_1}{r} + g_{2\,max} \cdot r_1.\ln\dfrac{R}{r_1}$

$$= 38 \times 1.5 \times \ln\frac{2.7}{1.5} + 30 \times 2.7 \times \ln\frac{4}{2.7}$$

$$= 65.841 \text{ KV}$$

Safe working voltage (rms) $= \dfrac{65.34}{\sqrt{2}} = 46.2$ KV

13. **A single core cable has a core diameter of $0.8\,cm$ and sheath diameter of $2\,cm$. The relative permittivity of the dielectric is 4. The power factor on open circuitis 0.03. Calculate for $1\,km$ length of the cable, (i) the capacitance (ii) charging current (iii) the dielectric loss when the cable is connected to $10\,KV$, $1-\phi 50\,Hz$ supply system? (iv) The equivalent insulation resistance?**

A. Given data:

Diameter of core, $d = 0.8$ cm

Sheath diameter, $D = 2$ cm

Relative permittivity, $\epsilon_r = 4$

Power factor, $\cos\phi = 0.03$

Length of cable, $l = 1$ km

Line voltage, $V = 10$ KV

$$V_{Ph} = \frac{10 \times 10^3}{\sqrt{3}} = 5773.50 \text{ V}$$

(i) Capacitance $C = 2\Pi\epsilon_0\epsilon_r \ln\left(\dfrac{D}{d}\right) \times 1000$ for 1 km. Length

$$= 2\Pi \times 8.854 \times 10^{-12} \times 4 \times \ln\left(\dfrac{2}{0.8}\right) \times 1000$$

$$C = 0.24\ \mu f$$

(ii) Charging current $I_C = 2\Pi fCV$

$$= 2\Pi \times 50 \times 0.24 \times 10^{-6} \times 5773.50$$

$$I_C = 0.435\ \text{Amps}$$

(iii) The dielectric loss, $P = WCV^2 \tan\delta$

$$\delta = \dfrac{\pi}{2} - \phi$$

$$\cos\phi = 0.03$$

$$\phi = \cos^{-1}(0.03) = 88.28°$$

$$\delta = 90° - 88.28° = 1.72°$$

$$P = 2\pi \times 50 \times 6.24 \times 10^{-6} \times (5773.50)^2 \tan(1.72)$$

$$P = 75.47\ \text{watts}$$

(iv) Equivalent insulation resistance $R = ?$

$$P = \dfrac{V^2}{R}$$

$$R = \dfrac{(5773.50)^2}{75.47}$$

$$R = 441.67\ M\Omega$$

CHAPTER 8

Distribution Systems

8.1 Introduction

The electrical energy produced at the generating station is conveyed to the consumers through a network of transmission and distribution systems. It is often difficult to draw a line between the transmission and distribution systems of a large power system. It is impossible to distinguish the two merely by their voltage because what was considered as a high voltage a few years ago is now considered as a low voltage. In general, distribution system is that part of power system which distributes power to the consumers for utilization.

8.2 Distribution System

That part of power system which distributes electric power for local use is known as distribution system.

In general, the distribution system is the electrical system between the sub-station fed by the transmission system and the consumers meters. It generally consists of feeders, distributors and the service mains.

Components of Distribution System

There are three components of distribution system

1. Feeder
2. Distributor
3. Service Mains

(1) Feeder:

A feeder is a conductor which connects the sub-station to the area where power is to be distributed. Generally, no tapings are taken from the feeder so that current in it remains the same throughout. The main consideration in the design of a feeder is current carrying capacity.

(2) Distributor:

A distributor is a conductor from which tapings are taken of supply to the consumers. In Figure shown below AB, BC, CD, DA are the distributors. The current through a distributor is not constant because tapings are taken at various places along its length. While designing a distributor, voltage drop along its length is the main consideration since the statutory limit of voltage variations is ±6% of rated value at the consumer's terminals.

(3) Service Mains:

A service main is generally a small cable which connects the distributor to the consumer's terminals.

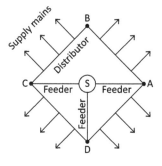

Fig. 8.1

8.3 Classification of Distribution Systems

A distribution system may be classified according to:

(i) Nature of Current:

According to nature of current, distribution system may be classified as

a. D.C distribution system
b. A.C distribution system

(ii) **Type of Construction:**

According to type of construction, distribution system may be classified as

a. Overhead System
b. Underground System

The Overhead System is generally employed for distribution as it is 5 to 10 times cheaper than the equivalent underground system.

(iii) **Scheme of Connection:**

According to scheme of connection, the distribution system may be classified as

a. Radial Main System
b. Ring Main System
c. Inter-Connected System

8.3.1 A.C Distribution

Now-a-days electrical energy is generated, transmitted and distributed in the form of alternating current. One important reason for the wide spread use of alternating current in preference to direct current is the fact that alternating voltage can be conveniently changed in magnitude by means of a transformer.

There is no definite line between transmission and distribution according to voltage (or) bulk capacity. However, in general, the a.c distribution system is the electrical system between the step-down substation fed by the transmission system and the consumer's meters. The a.c distribution system is classified into

i. Primary distribution system
ii. Secondary distribution system

(i) **Primary distribution system:**

It is that part of A.C distribution system which operates at voltages somewhat higher than general utilization and handles large blocks of electrical energy than the average low-voltage consumer uses. The voltage used for primary distribution depends upon the amount of power to be conveyed and the distance of the substation required to be fed. The mostly used primary distribution voltages are 11KV, 6.6KV and 3.3KV

ii) **Secondary distribution system:**

It is the part of a.c distribution system which includes the range of voltages at which the ultimate consumer utilizes the electrical energy delivered to him. The secondary distribution employs 400/230V, 3-phase, 4-wire system.

8.3.2 D.C Distribution

It is a common knowledge that electric power is almost exclusively generated, transmitted and distributed as a.c. However, for certain applications, d.c supply is absolutely necessary. For instance, d.c supply is required for the operation of variable speed machinery (i.e., dc motors), for electro chemical work and for congested areas where storage battery reserves are necessary.

The d.c supply from the substation may be obtained in the form of

 i. 2-wire distribution
 ii. 3-wire distribution

(i) **2-Wire D.C System:**

As the name implies, this system of distribution consists of two wires. One is the outgoing (or) positive wire and the other is the return (or) negative wire. The loads such as lamps, motors etc are connected in parallel between the two wires as shown in figure. This system is never used for transmission purposes due to low efficiency but may be employed for distribution of d.c power.

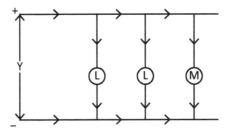

Fig. 8.2

(ii) 3-wire d.c system:

It consists of two outers and a middle (or) neutral wire which is earthed at the substation. The voltage between outers is twice the voltage between either outer and neutral wire as shown in figure. The principal advantage of this method is that it makes available two voltages at the consumer terminals viz., V between any outer and neutral and 2V between the outers. Loads requiring high voltage (e.g. motors) are connected across the outers, whereas lamps and heating circuits requiring less voltage are connected between either outer and the neutral.

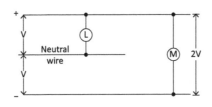

Fig. 8.3

8.4 Overhead versus Underground System

The distribution system can be overhead (or) underground.

(i) Public Safety:

The underground system is more safe than overhead system because all distribution wiring is placed underground and there are little chances of any hazard.

(ii) **Initial Cost:**

The underground system is more expensive due to high cost of trenching,

Conduits, cables, manholes and other special equipment. The initial cost of an underground system may be five to ten times than that of an overhead system.

(iii) **Faults:**

The chances of faults in underground system are very rare as the cables are laid underground and are generally provided with better insulation.

(iv) **Current Carrying Capacity and Voltage drop:**

An overhead distribution conductor has a considerably higher current carrying capacity than an underground cable conductor of the same material and cross-section. On the other hand, underground cable conductor has much lower inductive reactance than that of an overhead conductor because of closer spacing of conductors.

(v) **Maintenance Cost:**

The maintenance cost of underground system is very low as compared with that of overhead system because of fewer chances of faults and service interruptions from wind, ice, and lightning as well as from traffic hazards.

8.5 Connection Schemes of distribution System

8.5.1 Radial System

In this system, separate feeders radiate from a single substation and feed the distributors at one end only. Figure(i) shows a single line diagram of a radial system for d.c distribution where a feeder D.C supplies a distributor AB at point 'A', Figure (ii) shows a single line diagram of radial system for a.c distribution. The radial system is employed only when power is generated at low voltage and the substation is located at the centre of the load.

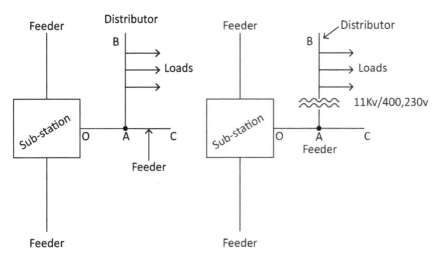

Fig. 8.4

Advantages:
- Construction is simple
- Low initial cost

Disadvantages:
- The end of the distributor nearest to the feeding point will be heavily loaded.
- The consumers are dependent on a single feeder and single distributor. Any fault on the feeder (or) distributor cuts off supply to the consumers who are on the side of fault away from the substation.

8.5.2 Ring Main System

In this system, the primaries of distribution transformers form a loop. The loop circuit starts from the substation bus-bars makes a loop through the area to be served and returns to the substation. The below figure shows the single line diagram of ring main system for d.c distribution where substation supplies to be closed feeder LMNOPQRS. The distributors are tapped from the different points M, O and Q of the feeder through distribution transformers. The ring main system has the following advantages:

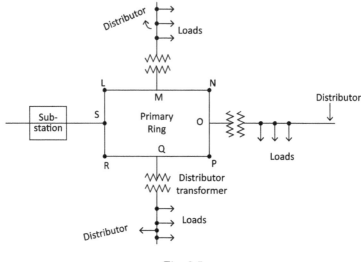

Fig. 8.5

i. There are less voltage fluctuations at consumer's terminals.
ii. The system is very reliable as each distributor is fed via "two feeders. In the event of fault on any section of the feeder, the continuity of supply is maintained.

8.5.3 Inter Connected System

When the feeder ring is energized by two (or) more than two generating stations (or) substations, it is called inter-connected system. Figure below shows the single line diagram of inter connected system where the closed feeder ring ABCD is supplied by two substations S1 and S2 at points D and C respectively. Distributors are connected to points O, P, Q and R of the feeder ring through distribution transformers. The inter connected system have the following advantages:

a. It increases the service reliability
b. Any area fed from one generating station during peak load hours can be fed from the other generating station. This reduces reserve power capacity and increases efficiency of the system.

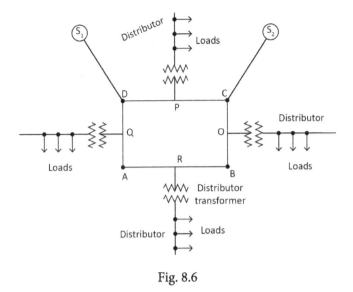

Fig. 8.6

8.6 Requirements of a Distribution System

(i) **Proper voltage:**

One important requirement of a distribution system is that voltage variations at consumer's terminals should be as low as possible. Low voltage causes loss of revenue, inefficient lighting and possible burning out of motors. High voltage causes lamps to burn out permanently and may cause failure of the other appliances. The statutory limit of voltage variations is ±6% of rated value at the consumer's terminals.

(ii) **Availability of Power on demand:**

Power must be available to the consumers in any amount that they may require from time to time. For example, motors may be started (or) shut down, lights may be turned on (or) off, without advance warning to the electric supply company. As electrical energy cannot be stored, therefore the distribution system may be capable of supplying load demands of the consumers.

(iii) **Reliability:**

Modern industry is almost dependent on electric power for its operation. Homes and office buildings are lighted, heated, cooled and ventilated by electric power. This calls for reliable service. The reliability can be improved to a considerable extent by:

a. Inter connected system
b. Reliable automatic control system
c. providing additional reserve facilities.

8.7 Types of D.C Distributors

The most general method of classifying d.c distributors is the way they are fed by the feeders.

i. Distributor fed at one end
ii. Distributor fed at both ends
iii. Distributor fed at the centre.

8.7.1 Distributor Fed at One End (Concentrated Loading)

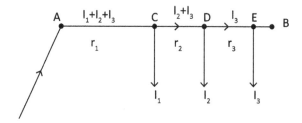

Fig. 8.7

In this type of feeding, the distributor is connected to the supply at one end and loads are taken at different points along the length of the distributor. The figure shows the single line diagram of a d.c distributor AB fed at the end A (also known as singly fed distributor) and loads I_1, I_2, I_3 tapped off at points C, D and E respectively.

let r_1, r_2, r_3 be the resistances of both wires (go and return) of the sections AC, CD, DE of the distributor respectively.

Current in section $AC = I_1 + I_2 + I_3$

Current in section $DC = I_2 + I_3$

Current in section $DE = I_3$

Voltage drop in section $AC = r_1(I_1 + I_2 + I_3)$

Voltage drop in section $CD = r_2(I_2 + I_3)$

Voltage drop in section $DE = r_3 I_3$

Therefore symbol Total voltage drop in the distributor

$= r_1(I_1 + I_2 + I_3) + r_2(I_2 + I_3) + I_3 r_3$

The following points are worth noting in a singly fed distributor:

a. The current in the various sections of the distributor away from the feeding point goes on decreasing. Thus current in section AC is more than current in section CD and current in section CD is more than current in section DE
b. The voltage across the loads away from the feeding point goes on decreasing. The figure shown above has the minimum voltage occurs at the load point E.
c. In case a fault occurs on any section of the distributor, the whole distributor will have to be disconnected from the supply mains. Therefore, continuity of supply is interrupted.

8.7.2 Distributor Fed at Both Ends

In this type of feeding, the distributor is connected to the supply mains at both ends and loads are tapped off at different points along the length of the distributor. The voltage at the feeding points may (or) may not be equal. The figure below shows a distributor AB fed at the ends A and B and loads of I_1, I_2, I_3 tapped off at pointsC, D and E. Here, the load voltage goes on decreasing as we move away from one feeding point say A, reaches minimum value and then again starts rising and reaches the maximum value when we reach other feeding point B. The minimum voltage occurs at some load point and is never fixed. It is shifted with the variation of load on different sections of distributor.

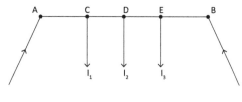

Fig. 8.8

Advantages:

a. If a fault occurs on any feeding point of the distributor, the continuity of supply is maintained from other feeding point.
b. In case of fault on any section of distributor, the continuity of supply is maintained from the other feeding point.

8.7.3 Distributor Fed at the Centre

In this type of feeding, the centre of the distributor is connected to the supply mains as shown in figure. It is equivalent to two singly fed distributors, each distributor having a common feeding point and length equal to half of the total length.

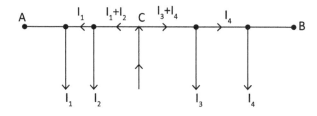

Fig. 8.9

Problems

1. A 2-wire DC distributor AB is 2km long and supplies a load of 100A, 150A, 200A, 50A situated at 500m, 1000m, 160m, 200m from the feeding point "A". Each conductor has a resistance of 0.01Ω /1000 m. Calculate the potential difference at each load, if voltage at Point A is 300V.

Sol: Given voltage at point A = 300V , $R = \dfrac{0.01\Omega}{1000m}$ (for conductor)

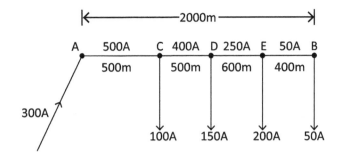

∴ Resistance of distributor per 1000m = 2 × 0.01

$$= 0.02\,\Omega$$

Resistance of section $AC = 0.02 \times \dfrac{500}{1000}$

$$= 0.01\,\Omega$$

Resistance of section $CD = 0.02 \times \dfrac{500}{1000} = 0.01\,\Omega$

Resistance of section $DE = 0.02 \times \dfrac{600}{1000} = 0.012\,\Omega$

Resistance of section $EB = 0.02 \times \dfrac{400}{1000} = 0.008\,\Omega$

∴ Potential difference at 'C' $= V_A - I_{AC} \cdot R_{AC}$

$$= 300 - (500)(0.01)$$

$$= 295\,V$$

Potential difference at 'D' $= V_C - I_{CD} \cdot R_{CD}$

$$= 295 - (400)(0.01)$$

$$= 291\,V$$

Potential difference at 'E' $= V_D - I_{ED} \cdot R_{ED}$

$$= 288\,V$$

Potential difference at 'B' $= V_E - I_{BE} \cdot R_{BE}$

$$= 288 - (50)(0.008)$$

$$= 287.6V$$

2. **A 2-wire DC distributor AB is 300m long and it is fed at A. the various loads and their positions are given below:**

At point	Distance from A	Concentrated load (A)
C	40	30
D	100	40
E	150	100
F	250	50

If the max. Permissible voltage drop will not exceeds $10V$, find the cross-sectional area of distributor. Consider $\rho = 1.78 \times 10^{-8} \Omega m$.

Sol:

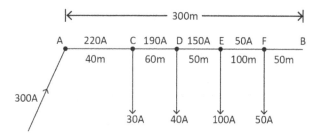

Given $\rho = 1.78 \times 10^{-8} \Omega$

Length of distributor $= 300m$

Let us assume that resistance of 100m length of the distributor is "r" Ω

Resistance of various sections are $R_{AC} = 0.4r\Omega$, $R_{DC} = 0.6r\Omega$, $R_{DE} = 0.5r\Omega$, $R_{EF} = r\Omega$

The currents in various sections of distributors are

$$I_{AC} = 220A, I_{CD} = 190A, I_{DE} = 150A, I_{EF} = 50A$$

∴ The total voltage drop of the distributor is

$$I_{AC}.R_{AC} + I_{DC}.R_{DC} + I_{DE}.R_{DE} + I_{EF}.R_{EF} = 10V$$

$$(220)(0.4r) + (190)(0.6r) + (150)(0.5r) + (50)(r) = 10$$

$$327r = 10$$

$$r = \frac{10}{327}$$

$$r = 0.0305$$

Cross-sectional area $A = \dfrac{\rho l}{A}$

$$= \frac{1.78 \times 10^{-8} \times 300}{0.0305}$$

$$A = 1.7508 \times 10^{-4} \, m^2$$

3. The load distribution of 2-wire distribution system is shown in figure.

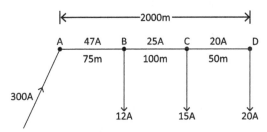

The cross-sectional area of the each conductor is $0.27 cm^2$. The end A is supplied at 250V. Resistivity of the wire is $1.78\mu\Omega m$. Calculate (i) Current in each section of conductor (ii) two core resistances each section (iii) voltage at each tapping point.

Sol: Total current = 47 A

(i)   Current in section $AB = 47 A$

         Current in section BC 25 A

         Current in section $CD = 20 A$

$A = 0.27 cm^2$, $\rho = 1.78 \mu\Omega cm$, $l = 225m$

(ii) $R = \dfrac{\rho l}{A}$

$R_{AB} = \dfrac{1.78 \times 10^{-6} \times 75 \times 10^{-2} \times 2}{0.27 \times 10^{-4}} = 9.88 \times 10^{-2}\,\Omega$

$R_{BC} = \dfrac{1.78 \times 10^{-6} \times 10^{-2} \times 100 \times 2}{0.27 \times 10^{-4}} = 13.18 \times 10^{-2}\,\Omega$

$R_{CD} = \dfrac{1.78 \times 10^{-6} \times 10^{-2} \times 50 \times 2}{0.27 \times 10^{-4}} = 6.58 \times 10^{-2}\,\Omega$

(iii) Voltage at $B = V_A - I_{AB} R_{AB}$

$= 250 - (37)(4.94 \times 10^{-2})$

$= 248.17\,V$

Voltage at $C = V_B - I_{BC} R_{BC}$

$= 248.17 - (25)(6.59 \times 10^{-2})$

$= 246.52\,V$

Voltage at $D = V_C - I_{CD} R_{CD}$

$= 246.52 - (20)(3.29 \times 10^{-2})$

$= 245.862\,V$

8.8 Distributor Fed at One End – Uniformly Distributed Load

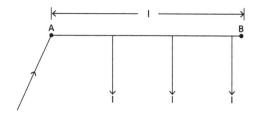

Fig. 8.10

The above figure shows the single line diagram of 2-wire system AB fed at "A" with uniformly distributed loads of "I" amp.

Let "l" be the length of the distributor AB and 'r' Ω be

The resistance per meter length.

Consider a point "C" at a distance of 'x' m from the feeding point 'A' as shown in fig (B)

Current at $C = I(1-x)$ Amp

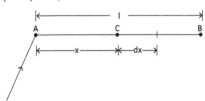

Now, consider a small length dx at "C" then its resistance will be r.dx

The voltage drop over the length dx is $dv = I(1-x)rdx$

∴ Total voltage drop in the distribution upto point C is

$$\int_0^x dV = \int_0^x i(1-x)rdx$$

$$V = ir\int_0^x (1-x)dx$$

$$= ir\left(lx - \frac{x^2}{2}\right)_0^x$$

$$V = \frac{I r x^2}{2}$$

The voltage drop up to point 'B' (total length of distribution) can be obtained by putting $x = 1$

Hence, above expression i.e.

$$V = ir\left(2x - \frac{x^2}{2}\right)$$

$$V = ir\left(1.1 - \frac{1^2}{2}\right)$$

$$V = \frac{irl^2}{2}$$

$$V = \frac{(il)(lr)}{2}$$

$$V = \frac{1}{2}IR$$

Where I = total current entering the feeding point 'A'.

R = Total resistance of the distribution

Thus, in uniformly loaded distributor fed at one end, the total voltage drop is equal to the produced by the whole of the load assumed to be concentrated at the middle point.

1. **A 2-wire DC distributor 200m long is uniformly loaded with 2A/m. Resistance of single wire is 0.3Ω/km. If the distributor is fed at one end. Calculate (i) voltage drop upto distance of 150m from the feeding point (ii) The max. voltage drop**

Sol: Given $l = 200m, i = 2A/m$,

$$r = 0.3\Omega/km = \frac{0.3}{1000} = 3 \times 10^{-4} m,$$

$x = 150m$

Resistance of the distributor $= 3 \times 10^{-4} \times 2 = 6 \times 10^{-4} m$

(i) Voltage drop upto 150m is $V = ir\left(lx - \frac{x^2}{2}\right)$

$$= 2 \times 6 \times 10^{-4} \left(200 \times 150 - \frac{(150)^2}{2}\right)$$

$V = 22.5 \text{volts}$

(ii) Max. voltage drop $= \frac{1}{2}(il)(rl)$

$$= \frac{1}{2}(2 \times 200)(6 \times 10^{-4} \times 200)$$

$$V = 24 \text{volts}$$

2. A 250m, 2-wire distributor fed from one end is loaded uniformly at the rate of 1.6A/m. The resistance of each conductor is 0.0002Ω/m0.0002Ω/m. Find the voltage necessary at feeding point to maintain 250V (i) at far end (ii) at mid point of distributor

Given $l = 250m$, $i = 1.6 A/m$, $r = 0.0002 \Omega/m$

Resistance of distributor $= 2 \times 0.0002 = 0.0004 \Omega/m$

$$I = il = 1.6 \times 250 = 400 A/m$$

$$R = rl = 0.0004 \times 250 = 0.1 \Omega/m$$

$$x = \frac{250}{2} = 125$$

Voltage $(V) = \frac{1}{2}(I)R$

$$= \frac{1}{2} \times 400 \times 0.1$$

$$V = 20 \text{volts}$$

(i) Voltage at far end $= 250 + 20$

$$= 270 V$$

(ii) Voltage at midpoint of distributor $= ir\left(lx - \frac{x^2}{2}\right)$

$$= 1.6 \times 0.0004 \left(250 \times 125 - \frac{(125)^2}{2}\right)$$

$$= 15 V$$

∴ Total voltage = 250 + 15

= 265V

8.9 Distributor Fed at Both Ends – Concentrated Loading

Whenever possible, it is desirable that a long distributor should be fed at both ends instead of at one end only, since total voltage drop can be considerably reduced without increasing the cross-section of the conductor. The two ends of the distributor may be supplied with (i) equal voltages (ii) unequal voltages.

8.9.1 Two Ends with Equal Voltages

Consider a distributor AB fed at both ends with equal voltage 'V' volts and having concentrated loads I_1, I_2, I_3, I_4 and I_5 at points C, D, E, F and G respectively as shown in figure. As we move away from one of the feeding points, say A potential difference (p.d) goes on decreasing till it reaches the minimum value at some load point, say E, and then again starts rising and becomes V volts as we reach the other feeding point B.

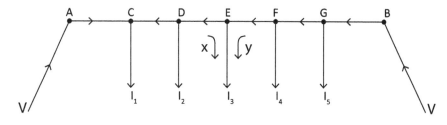

Fig. 8.11

All the currents tapped off between points A and E (minimum p.d point) will be supplied from feeding point A while those tapped off between B and E will be supplied from the feeding point B. The current tapped off at point E itself will be partly supplied from A and partly from B. If these currents are x and y, then

$$I_3 = x + y$$

∴ We arrive at a very important conclusion that at the point of minimum potential, current comes from both ends of distributor.

Point of minimum potential:

It is generally desired to locate the point of minimum potential. There is a simple method for it. Consider a distributor AB having three concentrated loads I1, I2 & I3 at points C, D, E. Suppose that current supplied by feeding end A is I_A, then current distribution in various sections of distributor can be worked out as shown in figure.

$$I_{AC} = I_A, \qquad I_{CD} = I_A - I_1$$

$$I_{DE} = I_A - I_1 - I_2, \qquad I_{EB} = I_A - I_1 - I_2 - I_3$$

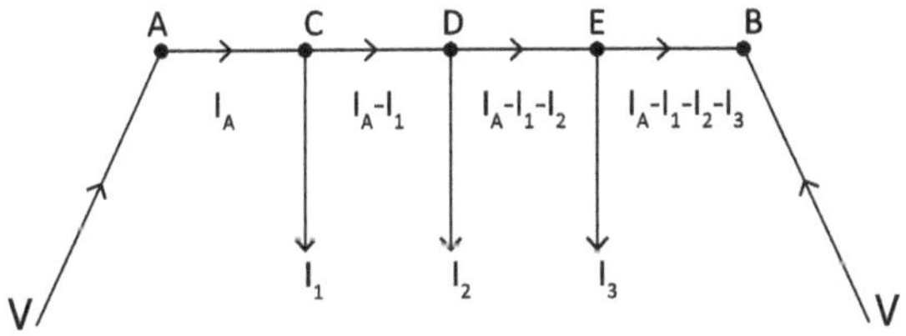

Fig. 8.12

Voltage drop between A and B = voltage drop over AB (or)

$$V - V = I_A R_{AC} + (I_A - I_1)R_{CD} + (I_A - I_1 - I_2)R_{DE} + (I_A - I_1 - I_2 - I_3)R_{EB}$$

From this equation, the unknown I_A can be calculated as the values of other quantities are given.

> The load point where the currents are coming from both sides of the distributor is the point of minimum potential.

8.9.2 Two Ends Fed with Unequal Voltages

The figure below shows the distributor AB fed with unequal voltages, end A being fed at V_1, volts and end B at V_2 volts. The point of minimum potential can be found by following the same procedure as discussed above. Thus in this case

Voltage drop between A & B = voltage drop over AB

$$V_1 - V_2 = \text{voltage drop over AB}$$

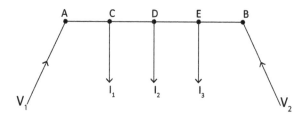

Fig. 8.13

1. **A 2-wire DC distributor AB 600M long is fed from both ends at 220V. Loads of 20A, 40A, 50A and 30A are tapped at a distance of 100M, 250M, 400M, 500M from the end A. If the area of cross-section of distribution conductor is 1cm². Find the minimum consumer voltage. Take $\rho = 1.7 \times 10^{-6} \Omega Cm$.**

Sol: Given $l = 600M$, $A = 1 \times 10^{-4} M$, $V_A = V_B = 220V$

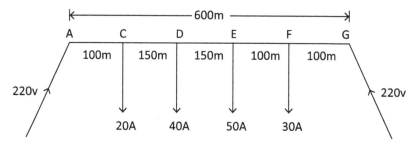

Current in section $AC = I_a$

Section $CD = I_a - 20$

Section $DE = I_a - (20 + 40) = I_a - 60$

Section $EF = I_a - (20 + 40 + 50) = I_a - 110$

Section $FB = I_a - (20 + 40 + 50 + 30) = I_a - 140$

∴ Resistance of '1m' length of the distributor is $R = \dfrac{\rho l}{A}$

$$R = \dfrac{1.7 \times 10^{-6} \times^{-2} \times 2}{1 \times 10^{-4}}$$

$$R = 3.4 \times 10^{-4} \Omega$$

Resistance for section AC, $R_{AC} = R \times 100$
$$= 3.4 \times 10^{-4} \times 100$$
$$= 0.034 \, \Omega$$

Resistance for section CD, $R_{CD} = R \times 150$
$$= 3.4 \times 10^{-4} \times 150$$
$$= 0.051 \, \Omega$$

Resistance for section DE, $R_{DE} = R \times 150$
$$= 3.4 \times 10^{-4} \times 150$$
$$= 0.051 \, \Omega$$

Resistance for section EF, $R_{EF} = R \times 100$
$$= 3.4 \times 10^{-4} \times 100$$
$$= 0.034 \, \Omega$$

Resistance for section FB, $R_{FB} = 3.4 \times 10^{-4} \times 100$
$$= 0.034 \, \Omega$$

Voltage at point 'B'

$$V_B = V_A - I_{AC}R_{AC} - (I_a - 20)R_{CD} - (I_a - 60)R_{DE} - (I_a - 110)R_{EF} - (I_a - 140)R_{FB}$$

$$220 = 220 - I_a(0.034) - (I_a - 20)(0.051) - (I_a - 60)(0.051) - (I_a - 110)(0.034) - (I_a - 140)(0.034)$$

$$220 - 220 = -I_a(0.034) - 0.17 I_a + 12.58$$

$$0.204 I_a = 12.58$$

$$I_a = 61.66 \, A$$

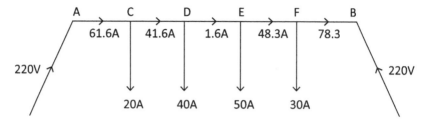

∴ Current in section AC = 61.66A

∴ Minimum potential point is at "E"

$$V_E = V_A - I_{ac}R_{ac} - I_{CD}R_{CD} - I_{DE}R_{DE}$$

$$= 220 - (61.6)(0.034) - (41.6)(0.051) - (1.6)(0.051)$$

$$= 220 - 2.0944 - 2.1216 - 0.816$$

$$V_E = 214.968 \text{ volts}$$

2. **A 2-wire DC distributor AB is fed from both ends. At feeding point 'A' the voltage is maintained as 230V & B at 235V. The total length of distributor is 200m and loads are 25A at 50m, 50A at 75m, 30A at 100m and 40A at 150 m from the point A. The resistance per km of one conductor is 0.3Ω. Calculate (i) current in various sections of distributor (ii) point of minimum voltage.**

Sol: Given $l = 200m$, $V_A = 230V$, $V_B = 235V$

```
|←――――――――――― 200m ―――――――――――→|
     A       C        D       E        F        G
            50m      25m     25m      50m      50m
220V ↗                                              ↖ 235V
             ↓        ↓       ↓        ↓
            25A      50A     30A      40A
```

Resistance of section AC, $R_{AC} = 0.3 \times 2 \times \dfrac{50}{1000} = 0.03\Omega$

Resistance of section CD, $R_{CD} = 0.3 \times 2 \times \dfrac{25}{1000} = 0.015\Omega$

Resistance of section DE. $R_{DE} = 0.3 \times 2 \times \dfrac{25}{1000} = 0.015\Omega$

Resistance of section EF, $R_{EF} = 0.03\Omega$

Resistance of section FB, $R_{FB} = 0.03\Omega$

Voltage at $V_B = V_A - I_{AC}R_{AC} - I_{CD}R_{CD} - I_{DE}R_{DE} - I_{EF}R_{EF} - I_{FB}R_{FB}$

$$235 = 230 - I_a(0.03) - (I_a - 25)(0.015) - (I_a - 75)(0.015)$$
$$- (I_a - 105)(0.03) - (I_a - 145)(0.03)$$

$$5 = -0.12I_a + 0.375 + 1.125 + 3.15 + 4.35$$

$$0.12I_a = 4$$

$$I_a = 33.33A$$

Current in section AC = 33.33A

Section CD = 33.33 − 25 = 8.33A

Section DE = −41.67A

Section EF = −71.67A

Section FB = −111.67A

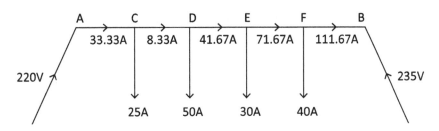

Minimum potential is at 'D'

$$\therefore V_D = V_A - I_{AC}R_{AC} - I_{CD}R_{CD}$$
$$= 230 - (33.33)(0.03) - (0.015)(8.33)$$
$$V_D = 228.87 \text{ V}$$

8.10 Distributor Fed at Both Ends – Uniform Loading

8.10.1 Distributor Fed at Both Ends with Equal Voltages

Consider a distributor AB of length 'l' meters, having resistance 'r' ohms per meter run and with uniform loading of i Amperes per meter run as shown in figure. Let the distributor be fed at feeding points A and B at equal voltages, say V volts. The total current supplied to the distributor is "il". As the two end voltages are equal, current supplied from each feeding point is $\frac{il}{2}$.

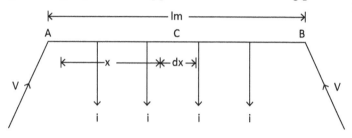

Fig. 8.14

Consider a point 'C' at a distance of 'x' meters from the feeding point A. Then current at point is $= i\frac{1}{2} - ix$

$$= i\left(\frac{1}{2} - x\right)$$

Now, consider small length 'dx' near point 'C'. it's resistance is rdx, voltage drop over the length 'dx' is

$$dV = i\left(\frac{1}{2} - x\right)rdx$$

∴ voltage drop up to point C $= \int_0^x dV = \int_0^x i\left(\frac{1}{2} - x\right)rdx$

$$V = ir\int_0^x \left(\frac{1}{2} - x\right)dx$$

$$V = ir\left(\frac{lx}{2} - \frac{x^2}{2}\right)$$

$$V = \frac{lr}{2}(lx - x^2)$$

Here the point of minimum potential will be the mid-point.

∴ Maximum voltage drop will occur at mid-point i.e., $x = \frac{1}{2}$

∴ Max. Voltage drop $= \dfrac{ir}{2}\left(1\left(\dfrac{1}{2}\right)-\dfrac{l^2}{4}\right)$

$= \dfrac{ir}{2}\left(\dfrac{l^2}{2}-\dfrac{l^2}{4}\right)$

$= \dfrac{ir}{2}\left(\dfrac{l^2}{4}\right)$

$= \dfrac{(il)(rl)}{8}$

$V = \dfrac{IR}{8}$

∴ Minimum voltage $= V - \dfrac{IR}{8}$

8.10.2 Distributor Fed at Both Ends with Unequal Voltage

Consider a distributor AB of length 'l' meters having resistance r ohms/meter with a uniform loading of i amp/meter as shown in figure. Let the distributor be fed from feeding points A and B at voltages V_A and V_B.

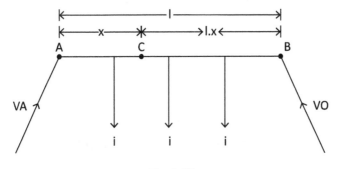

Fig. 8.15

Suppose that minimum potential "C" is situated at a distance of 'x' meters from the feeding point A. The current

Supplied by the feeding point 'A' will be i x.

Voltage drop in section AC $= \dfrac{irx^2}{2}$ volts

As the distance of C from feeding point B is (l − x)

∴ Current drop in section BC = $\dfrac{ir(l-x)^2}{2}$ volts

Voltage at point C, $V_C = V_A - \dfrac{ir(l-x)^2}{2}$ → (1)

Voltage at point C, $V_C = V_B - \dfrac{ir(l-x)^2}{2}$ → (2)

From equations (1) & (2)

$$V_A - \dfrac{irx^2}{2} = V_B - \dfrac{ir(l-x)^2}{2}$$

$$V_A - V_B = \dfrac{irx^2}{2} - \dfrac{ir(l-x)^2}{2}$$

$$= \dfrac{ir}{2}\left[x^2 - l^2 - x^2 + 2lx\right]$$

$$= \dfrac{ir}{2}\left[2lx - l^2\right]$$

$$2(V_A - V_B) = irl(2x - l)$$

$$\dfrac{2(V_A - V_B)}{irl} + l = 2x$$

$$x = \dfrac{2(V_A - V_B)}{irl} + l$$

As all the quantities on the right hand side of the equation are known, the point on the distributor where minimum potential occurs can be calculated.

1. A 2-wire DC distributor 100 m long is loaded with 0.5 A/meter. Resistance of each conductor is $0.05\Omega/Km$. Calculate the Max. voltage drop if the distributor is fed from both ends with equal voltages of 220 V. What is the minimum voltage & where it occurs.

Given $l = 100m$, $i = 0.5A/meter$, $R = 0.05\Omega/Km$, $V_1 = V_2 = 220V$

∴ Resistance of distributor $= 2 \times 0.05 = 0.1\Omega/km$

(i) Max. voltage drop $= \dfrac{IR}{8}$

$$= \dfrac{0.5 \times 100 \times 0.1 \times \dfrac{100}{1000}}{8} = 8$$

(ii) The minimum potential occurs at mid point

Minimum voltate = V − Max. Voltage drop

$= 220 - 0.0625$

$= 219.93$ V

2. **A 2-wire DC distributor AB 500 m long is fed from both ends and is loaded uniformly at rate of 1 A/meter. At feeding point 'A' the voltage is at 255 V and B is 250 V. If the resistance of each conductor is 0.1 Ω/Km. Calculate (i) Min. Voltage & Point where it occurs (ii) currents supplied from feeding point A and B.**

Given $V_A = 255$ A, $V_B = 250$ V, $l = 500$ m, $i = 1$ A/meter

Resistance of each conductor $= 0.1 \Omega / Km$

Resistance of distributor $r = 2 \times 0.1 \Omega / Km$

(i) Min potential at C $= \dfrac{V_A - V_B}{irl} + \dfrac{l}{2}$

$$= \dfrac{255 - 250}{1 \times 0.2 \times \dfrac{500}{1000}} + \dfrac{500}{2}$$

$= \dfrac{5}{0.1} + 250$

$= 50 + 250$

$= 300$ m

∴ Min. Voltage occurs at 300 m from point A

$$V_C = V_A - \dfrac{irx^2}{2}$$

$$= 255 - \left(1 \times 0.2 \times \frac{1}{100} \times \frac{(300)^2}{2}\right)$$

$$= 250 - \frac{18000}{1000 \times 2}$$

$$= 246 \text{ V}$$

Current supplied from feeding point (A) $= ix = 1 \times 300 = 300$ A

Current supplied from feeding point (B) $= i(l-x)$

$$= 1(500-300)$$

$$= 200 \text{ A}$$

Ring Main distribution system:

A distributor arranged to form a closed loop and fed at one (or) more points is called ring distributor. Such a distributor starts from one point, makes a loop through the area to be served, and returns to the original point. For the purpose of calculating voltagedistribution. The distributor can be considered as consisting of a series of open distributors fed at both ends. The principal advantage of ring distributor is that by proper choice in the number of feeding points, great economy in copper can be affected.

The simplest case of a ring distributor is the one having only one feeding point.

3. A 2-wire DC ring distributor is 300 m long and is fed at 240 V at point 'A'. At point 'B' 150 m from A, a load of 120 A is taken and at 'C' 100 m in the opposite direction, a load of 80 A is taken. If the resistance per 100 m of single Conductor is 0.03Ω. Find (i) Current in each section of the distributor (ii) voltage at point B and C.

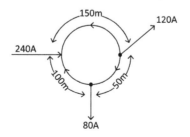

Resistance of conductor $= 0.03\,\Omega/100$ m

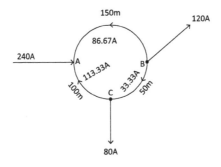

Resistance of distributor $= 0.03 \times 2$
$$= 0.06\,\Omega/100 \text{ m}$$

Resistance of section $AB = 0.06 \times \dfrac{150}{100}$
$$= 0.09\,\Omega$$

Resistance of section $BC = 0.06 \times \dfrac{50}{100}$
$$= 0.03\,\Omega$$

Resistance of section $CA = 0.06 \times \dfrac{100}{100}$
$$= 0.06\,\Omega$$

According to KVL
$$O = I_A R_{AB} + (I_A - 120)R_{BC} + (I_A - 200)R_{CA}$$
$$= I_A(0.09) + (I_A - 120)(0.03) + (I_A - 200)(0.06)$$
$$O = 0.18 I_A - 15.6$$
$$I_A = 86.67 \text{ A}$$

Current in section $AB = 86.67$ A
$$BC = -33.33 \text{ A}$$
$$CA = -113.33 \text{ A}$$

Voltage at Point B is
$$V_B = V_A - I_A R_{AB}$$
$$V_B = 232.19 \text{ V}$$

Voltage at point 'C' is

$$V_C = V_A - I_{AC}.R_{AC}$$
$$= 240 - (113.33)(0.06)$$
$$V_C = 233.2 \text{ V}$$

4. A 2-wire DC distributor ABCDEA in the form of a ring main is fed at point A at 220 V, and is loaded as under 10 A at B, 20 A at C, 30 A at D, 10 A at E. The resistance of various sections (go and return) are $AB = 0.1\Omega$, $BC = 0.05\Omega$, $CD = 0.01\Omega$, $DE = 0.025\Omega$, $EA = 0.075\Omega$. Determine (i) The point of min. Potential (ii) current in each section of distributor.

$V_A = 220$ V
Resistance of section $AB = 0.1\Omega$
$BC = 0.05\Omega$
$CD = 0.05\Omega$
$DE = 0.025\Omega$
$EA = 0.075\Omega$

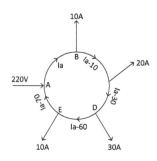

By applying KVL

$$0 = 0.1I_a + (I_a - 10)(0.05) + (I_a - 30)(0.01) + (I_a 60)(0.025) + (I_a - 70)(0.075)$$
$$0 = 0.26I_a - 7.55$$

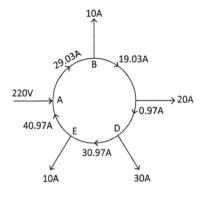

$$0.26 I_a = 7.55$$

$$I_a = 29.03 \text{ A}$$

(ii) Current in section $AB = 29.03$ A

$BC = 19.03$ A

$CD = -0.97$ A

$DE = -30.97$ A

$EA = -40.97$ A

(i) Minimum Potential point is "C".

8.11 A.C Distribution System

Method of solving A.C distribution problem:

In A.C distribution system, power factors of various load currents have to be considered since currents in different sections of the distributor will be the vector sum of load currents and not the arithmetic sum. The power factors of load currents may be given (i) w.r.t receiving (or) sending end voltage (ii) w.r.t load voltage itself.

8.11.1 Power Factor Referred to Receiving End Voltage

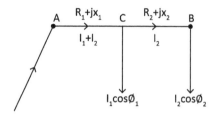

Consider an a.c distributor AB with concentrated loads of I_1 and I_2 tapped off at points C and B as shown in figure. Taking the receiving end voltage V_B as the reference vector, let lagging power factors at C and B be $cos\phi_1$ and $cos\phi_2$ w.r.t V_B. Let R_1, x_1 and R_2, x_2 be the resistance and reactance of sections AC and CB of the distributor.

Impedance of the section $AC, \overline{Z_{AC}} = R_1 + jx_1$

Impedance of the section $BC, \overline{Z_{BC}} = R_2 + jx_2$

Load current at point C, $\overline{I_1} = I_1(cos\phi_1 - jsin\phi_1)$

Load current at point B, $\overline{I_2} = I_2(cos\phi_2 - jsin\phi_2)$

Current in point $CB = \overline{I_{CB}} = \overline{I_2} = I_2(cos\phi_2 - jsin\phi_2)$

Current in point $AC = \overline{I_{AC}} = \overline{I_1} + \overline{I_2}$

$\qquad = I_1(cos\phi_1 - jsin\phi_1) + I_2(cos\phi_2 - jsin\phi_2)$

Voltage drop in section CB, $\overline{V_{CB}} = \overline{I_{CB}} \overline{Z_{CB}}$

$\qquad I_2(cos\phi_2 - jsin\phi_2)(R_2 + jx_2)$

Voltage drop in section AC, $\overline{V_{AC}} = \overline{I_{AC}} \overline{Z_{AC}}$

$\qquad = [I_1(cos\phi_1 - jsin\phi_1) + I_2(cos\phi_2 - jsin\phi_2)](R_1 + jx_1)$

Sending end voltage $\overline{V_A} = \overline{V_{AC}} + \overline{V_{AB}} + \overline{V_B}$

Sending end current $\overline{I_A} = \overline{I_1} + \overline{I_2}$

Phasor diagram:

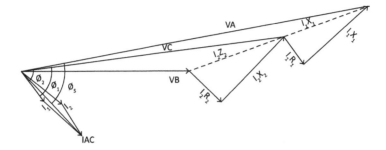

8.11.2 Power Factors Referred to Respective Load Voltages

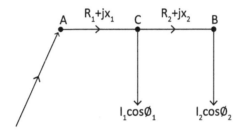

The power factors of loads are referred to their respective load voltages. Then ϕ_1 is the phase angle between V_C and I_1, ϕ_2 is the phase angle between and I_2. The vector diagram under these conditions is shown in below fig.

Voltage drop in section $\overline{CB} = \overline{I_2} Z_{CB}$

$$= I_2 (cos\phi_2 - jsin\phi_2)(R_2 + jx_2)$$

Voltage at point $C = \overline{V_B}$ + Drop in section CB

Now, $\overline{I_1} = I_1 \angle -\phi_1$ w.r.t voltage V_C

$\therefore \overline{I_1} = I_1 \angle -(\phi_1 - \alpha)$ w.r.t voltage V_B.

$\overline{I_1} = I_1 \left[cos(\phi_1 - \alpha) - jsin(\phi_1 - \alpha) \right]$

Now, $\overline{I_{AC}} = \overline{I_1} + \overline{I_2}$

$$= I_1 \left[cos(\phi_1 - \alpha) - jsin(\phi_1 - \alpha) \right] + I_2 (cos\phi_2 - jsin\phi_2)$$

Voltage drop section $AC = \overline{I_{AC} Z_{AC}}$

Voltage at point $A = V_B$ + Drop in CB + Drop in AC

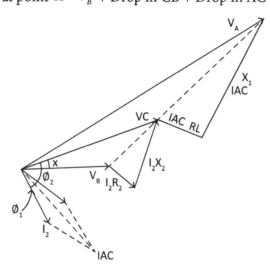

1. A single phase distributor 1 Km long has resistance and reactance per conductor of 0.1Ω and 0.15Ω respectively. At the far end, the voltage $V_B = 200$ V and the current is 100 A at a P.f of 0.8 lagging. At the mid point 'C' of the distributor a current of 100 A is tapping at a P.f. of 0.6 lagging with reference to the voltage V_C at the mid-point. Calculate:

(a) Voltage at mid-point
(b) Sending end voltage V_A
(c) Phase angle between V_A and V_B

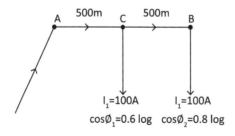

AB is the distributor with 'M' as mid-point

Total impedance of the distributor $= 2(0.1 + j0.15)$

$= (0.2 + j0.3)\Omega$

Impedance of section AC, $\overline{Z_{AC}} = (0.1 + j0.15)\Omega$
Impedance of section CB, $\overline{Z_{CB}} = (0.1 + j0.15)\Omega$

Let the voltage V_B at point 'B' be taken as the reference vector.

then, $\overline{V_B} = 200 + j0$

(a) Load current at point B, $\overline{I_2} = 100(0.8 - j0.6)$

$$= 80 - j60$$

Current section CB, $\overline{I_{CB}} = \overline{I_2} = 80 - j60$

Drop in section $\overline{V_{CB}} = \overline{I_{CB}} \cdot \overline{Z_{CB}}$

$$= (80 - j60)(0.1 + j0.15)$$

$$= 17 + j16$$

Voltage at point C, $\overline{V_C} = \overline{V_B} + \overline{V_{CB}}$

$$= (200 + j0) + (17 + j6)$$

$$= 217 + j6$$

$$= \sqrt{(217)^2 + (6)^2}$$

$$= 217.1 \text{ volts}$$

Phase angle between V_C and $V_B \alpha = \tan^{-1}\left(\dfrac{6}{217}\right)$

$$= \tan^{-1} 0.0276$$

$$= 1.58^0$$

(b) The load current I_1 has a lagging P.F of 0.6 w.r.t. V_C It lags behind V_C by an angle $\phi_1 = \cos^{-1} 0.6 = 53.13^0$

∴ Phase angle between I_1 and V_B, $\phi_1^| = \phi_1 - \alpha$

$$= 53.13^0 - 1.58^0$$

$$= 51.55^0$$

Load current at C, $\bar{I}_1 = I_1(\cos\phi_1^! - j\sin\phi_1^!)$

$\qquad = 100(\cos 51.55° - j\sin 51.55°)$

$\qquad = 62.2 - j78.3$

Current in section AC, $\bar{I}_{AC} = \bar{I}_1 + \bar{I}_2$

$\qquad = (62.2 - j78.3) + (80 - j60)$

$\qquad = 142.2 - j138.3$

Drop in section AC, $\bar{V}_{AC} = \bar{I}_{AC} \cdot \bar{Z}_{AC}$

$\qquad = (217 + j6) + (34.96 + j7.5)$

$\qquad = 251.96 + j13.5$

$\qquad = \sqrt{(251.96)^2 + (13.5)^2}$

$\qquad = 252.32 \text{ volts}$

(c) The phase difference 'θ' between V_A and V_B is given by

$$\tan\theta = \frac{13.5}{251.96}$$

$\qquad = 0.05358$

$\theta = \tan^{-1} 0.05358$

$\qquad = 3.07°$

Hence supply voltage is 252.32 volts and leads V_B by 3.07°.

2. **A 1-ϕ AC distributor AB 300 m long fed from A and is loaded as follows 100 A at 0.707 Pf lagging at 200 m from A. 200 A at 0.8 Pf lagging 300 m from point A. The load reactance f resistance of distributor is 0.1Ω & 0.2Ω per Km. calculate the total voltage drop in distributor. The power factors (P.f) are refer to the voltage at the far end.**

Impedance of distributor $= (0.2 + j0.1)\Omega/km$

Impedance of section AC, $\bar{Z}_{AC} = (0.2 + j0.1) \times \dfrac{100}{1000}$

$\qquad = (0.02 + j0.01)\Omega$

Load current at C, $\bar{I}_1 = I_1(\cos\phi_1 - j\sin\phi_1)$

$= 100(0.707 - j0.707)$

$= (70.7 - j70.7)$ A

Load current at B, $\bar{I}_2 = I_2(\cos\phi_2 - j\sin\phi_2)$

$= 200(0.8 - j0.6)$

$= (160 - j120)$ A

Current in section CB = $\bar{I}_2 = (160 - j120)$ A

Current in section AC $= \bar{I}_1 + \bar{I}_2$

$= (70.7 - j70.7) + 160 - j120)$

$= (230.7 - j190.7)$ A

Voltage drop in section CB $\bar{V}_{CB} = \bar{I}_{CB}\bar{Z}_{CB}$

$= (160 - j120)(0.02 + j0.01)$

$= (4.4 - j0.8)$ V

Voltage drop in section AC, $\bar{V}_{AC} = \bar{I}_{AC}\bar{Z}_{AC}$

$= (230.7 - j3.01) + (4.4 - j0.8)$

$= (17.44 - j3.81)$ V

In magnitude $= \sqrt{(17.44)^2 + (3.81)^2}$

$= 17.85$ V

3. A 1-ϕ AC distributor 2 km long supples a load of 120 A at 0.8 P.f lagging at its far end and a load of 80 A at 0.9 P.f lagging at its midpoint. Both P.f's are preferred to voltage at far end. The resistance and reactance per KM (go & return) are 0.05Ω & $= 0.01\Omega$. If the

voltage at far end is maintained at 230V. Calculate (i) voltage at sending end (ii) Phase angle b/w two end voltages.

Resistance of distributor $= 0.05\Omega$

Reactance of distributor $= 0.1\Omega$

Impedance of section AC, $\overline{Z_{AC}} = (0.05 + j0.1)\Omega$

Impedance of section BC, $\overline{Z_{BC}} = (0.05 + j0.1)\Omega$

Load current at C, $\overline{I_1} = I_1(cos\phi_1 - jsin\phi_1)$

$= 80(0.9 - j0.4)$

$= (72 - j32)$ A

Load current at B, $\overline{I_2} = i_2(cos\phi_2 - jsin\phi_2)$

$= 120(0.8 - j0.6)$

$= (96 - j72)$ A

Current in section CB, $\overline{I_2} = (96 - j72)$ A

Current in section $AC = \overline{I_1} + \overline{I_2}$

$= (72 - j32) + (96 - j72)$

$= (168 - j104)$ A

Voltage drop in section CB, $\overline{V_{CB}} = \overline{I_{CB}}\overline{Z_{CB}}$

$= (9 - j72)(0.05 + j0.1)$

$= (12 + j6)$ V

Voltage drop in section AC, $\overline{V_{AC}} = \overline{I_{AC}}\overline{Z_{AC}}$

$= (168 - J104)(0.05 + j0.1)$

$= (18.8 + 11.6j)$ V

(i) sending end voltage $= \overline{V_{AC}} + \overline{V_{BC}} + \overline{V_B}$

$= (18.8 + 11.6j) + (12 + 6j) + 230$

$= (260.8 + 17.6j)$ V

$= 261.3 \angle 3.86$ V

(ii) Phase angle b/w two end voltages

$$Tan\theta = \frac{V_A}{V_B}$$

$$= \frac{\angle 3.86}{\angle 0}$$

$Tan\theta = 3.86$

8.12 Stepped (or) Tapped Distributor

It is necessary to design each distributor with minimum volume of conductor material by satisfying the voltage limits.

Loads when tapped from any distributor changes the current along the length of distributor. For example, a uniformly loaded distributor supplying voltage at one end carries a current that changes from maximum value at feeding point to minimum at far end. If the uniform cross-section conductor is used throughout the distributor, it effects savings of conductor material. If the conductor cross-section is determined based on current to be carried i.e., if conductor is stepped, the required conductor material is minimum. When the conductor is stepped, the required conductor material is minimum. When the conductor is tapered then the cross-section of conductor at any point is proportional to the square root of distance from the far end of distributor, practically it is not possible.

Consider distributor AB with loads i_1, i_2 tapered at point C & D. Let l_1, a_1 & l_2, a_2 be the lengths and are of cross-sections of AC, CB.

Resistance of section AC, $R_{AC} = \frac{\rho l_1}{a_1}$

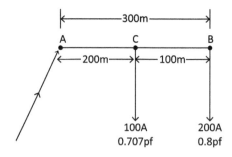

Resistance of section CB, $R_{CB} = \dfrac{\rho l_2}{a_2}$

Voltage drop in section AC, $V_{AC} = V_1 = (i_1 + i_2)R_{AC}$

$$V_1 = 2\dfrac{\rho l_1}{a_1}(i_1 + i_2)$$

$$a_1 = 2\dfrac{\rho l_1}{V_1}(i_1 + i_2)$$

Voltage drop in section CB, $V_{BC} = V - V_1 = RC_{CB}i_2$

$$= 2\dfrac{\rho l_2}{a_2}i_2$$

$$a_2 = 2\dfrac{\rho l_2}{V - V_1}i_2 \quad \to (2)$$

Volume of conductor material in the distributor can be written as

$$v = 2l_1 a_1 + 2l_2 a_2$$

$$v = 2l_1[(i_1 + i_2)\dfrac{2\rho l_1}{V_1} + 2l_1\left(\dfrac{2\rho l_2}{(V - V_1)}i_2\right)]$$

$$v = \dfrac{4\rho l_1^2}{V_1}(i_1 + i_2) + \dfrac{4\rho l_2^2}{(V - V_1)}i_2$$

Differentiate above equation w.r.t V_1

$$\dfrac{dv}{dV_1} = \dfrac{d}{dV_1}\left(\dfrac{4\rho l_1^2}{V_1}\right) + \dfrac{d}{dV_1}\left(\dfrac{4\rho l_2^2}{(V - V_1)}\right)$$

$$4\rho l_1^2(i_1 + i_2)\left(\dfrac{-1}{V_1^2}\right) + 4\rho l_2^2 i_2\left(\dfrac{1}{(V - V_1^2)}\right)$$

Substituting values of $V_1 V - V_1$ in the above equation and we can get the condition for minimum value of conductor material as

$$0 = \frac{-4\rho l_1^2}{V_1^2}(i_1 + i_2) + \frac{4\rho l_2^2 i_2}{(V-V_1)^2}$$

$$\frac{4\rho l_1^2 (i_1 + i_2)}{V_1^2} = \frac{4\rho l_2^2 i_2}{(V-V_1)^2}$$

$$\frac{l_1^2 (i_1 + i_2)}{V_1^2} = \frac{l_2^2 i_2}{(V-V_1)^2}$$

$$\frac{l_1^2 (i_1 + i_2) \times a^2}{(i_1 + i_2)^2 \cdot 4\rho^2 l_1^2} = \frac{l_2^2 i_2 a_2^2}{i_2^2 \cdot 4\rho^2 l_2^2}$$

$$\frac{a_1^2}{(i_1 + i_2) \cdot 4\rho^2 l_1^2} = \frac{a_2^2}{i_2 \cdot 4\rho^2}$$

$$\frac{a_1^2}{i_1 + i_2} = \frac{a_2^2}{i_2}$$

$$\frac{a_1}{a_2} = \sqrt{\frac{(l_1 + l_2)}{i_2}}$$

Objective Questions

1. Which of the following is usually not the generating Voltage?
   (a) 6.6 kV
   (b) 9.9 kV
   (c) 11kV
   (d) 13.2 kV
   Ans: b.

2. In overhead transmission lines the effect of capacitance can be neglected when the length of line is less than
   (a) 200 km
   (b) 160 km
   (c) 100 km
   (d) 80 km
   Ans: d.

3. In the analysis of short transmission lines which of the following is Neglected?
   (a) $I^2 R$ loss
   (b) Shunt admittance
   (c) Series impedance
   (d) All of the above
   Ans: b.

4. Surge impedance of transmission line is given by
   (a) $(L/C)^{1/2}$
   (b) $(C/L)^{1/2}$
   (c) $(CL)^{1/2}$
   (d) $1/(CL)^{1/2}$
   Ans: a.

5. 750 kV is termed as
   (a) Medium high voltage
   (b) High voltage
   (c) Extra high voltage
   (d) Ultra high voltage.
   Ans: d.

6. In case of transmission line conductors with the increase in atmospheric temperature
   (a) Length increase but stress decreases
   (b) Length increases and stress also increases
   (c) Length decreases but stress increases
   (d) Both length as well as stress decreases
   Ans: d.

7. If the height of transmission towers is increased, which of the following parameters is likely to change?
   (a) Resistance
   (b) Inductance
   (c) Capacitance
   (d) None of the above.
   Ans: c.

8. In medium transmission lines the shunt capacitance is taken into account in
   (a) Tee method
   (b) Pie method
   (c) Steinmetz method
   (d) all of the above.
   Ans: d.

9. In order to increase the limit of distance of transmission line
   (a) Series resistances are used
   (b) Synchronous condensers are used
   (c) Shunt capacitors and series reactors are used
   (d) Series capacitors and shunt reactors are used.
   Ans: d.

10. A 30 km transmission line carrying power at 33 kV is known as
    (a) Short transmission line
    (b) Long transmission line
    (c) High power line
    (d) Ultra high voltage line.

Ans: a.

11. The relation between travelling voltage wave and current wave is
    (a) $e = i (L/C)^{1/2}$
    (b) $e = i (C/L)^{1/2}$
    (c) $e = i (iL/C)^{1/2}$
    (d) $(L/iC)^{1/2}$

Ans: a.

12. Steepness of the traveling waves is attenuated by
    (a) Resistance of the line
    (b) Inductance of the line
    (c) Capacitance of the line
    (d) all of the above.

Ans: a.

13. The protection against direct lightening strokes and high voltage steep waves is provided by
    (a) Earthing of neutral
    (b) Lightening arresters
    (c) Ground wires
    (d) Lightening arresters and ground wires.

Ans: d.

14. The fact that a conductor carries more current on the surface as compared to core, is known as
    (a) Skin effect
    (b) Corona
    (c) Permeability
    (d) Unsymmetrical fault.

Ans: a.

15. Skin effect results in
    (a) Reduced effective resistance but increased effective internal reactance of the conductor
    (b) Increased effective resistance but reduced effective internal reactance of the conductor
    (c) Reduced effective resistance as well as effective internal reactance
    (d) Increased effective resistance as well as effective internal reactance.
Ans: b.

16. Skin effect depends on
    (a) Size of the conductor
    (b) Frequency of the current
    (c) Resistivity of the conductor material
    (d) All of the above.
Ans: d.

17. The skin effect of a conductor will reduce as the
    (a) Diameter increases
    (b) Frequency increases
    (c) Permeability of conductor material increases
    (d) Resistivity of conductor material increases.
Ans: d.

18. Skin effect is proportional to
    (a) diameter of conductor
    (b) (diameter of conductor)$^{1/2}$
    (c) (diameter of conductor)^2
    (d) (diameter of conductor)^2.
Ans: c.

19. Corona is accompanied by
    (a) Voilet visible discharge in darkness
    (b) Vibration
    (c) Hissing sound
    (d) All of the above
Ans: d.

20. Corona loss in a transmission line is dependent on
    (a) Diameter of the conductor
    (b) Material of the conductor
    (c) Height of the conductor
    (d) All the above

Ans: a.

21. The effect of corona is
    (a) Increased energy loss
    (b) Increased reactance
    (c) Increased inductance
    (d) All of the above

Ans: a.

22. Corona can be reduced by
    (a) increasing the operating voltage
    (b) reducing the space between conductors
    (c) increasing the effective conductor diameter
    (d) all of the above

Ans: c.

23. The charging current in a transmission line increases due to corona effect because corona increases
    (a) line current
    (b) effective line voltage
    (c) power loss in lines
    (d) the effective conductor diameter

Ans: d.

24. The chances of occurrence of corona are maximum during
    (a) humid weather
    (b) dry weather
    (c) winter
    (d) hot summer

Ans: a.

25. Corona is likely to occur maximum in case of
    (a) distribution lines
    (b) transmission lines
    (c) domestic wiring
    (d) service mains
    Ans: b.

26. The dielectric strength of air is
    (a) proportional to barometric pressure
    (b) proportional to absolute temperature
    (c) inversely proportional to barometric pressure
    (d) none of the above
    Ans: a.

27. Disruptive corona begins in smooth cylindrical conductors in air at NTPb. If the electric field intensity at the conductor surface goes upto
    (a) 21.1 kvrms/cm
    (b) 21.1 kv peak/cm
    (c) 21.1 kv average/cm
    (d) 21.1 kvrms/m
    Ans: a.

28. Power loss due to corona is directly proportional to
    (a) spacing between conductors
    (b) radius of conductor
    (c) supply frequency
    (d) none of the above
    Ans: c.

29. Corona losses are minimized when
    (a) conductor size is reduced
    (b) smooth conductor is used
    (c) sharp points are provided in the line hardware
    (d) current density in conductors is reduced
    Ans: b.

30. The corona loss on a particular system at 50 Hz is 1 kw/km per phase. The corona loss at 60 Hz would be
    (a) 1 kw/km per phase
    (b) 0.83 kw/km per phase
    (c) 1.2 kw/km per phase
    (d) 1.13 kw/km per phase
Ans: c.

31. Skin effect in a transmission line is due to
    (a) supply frequency
    (b) self inductance of conductor
    (c) high sensitivity of material in the centre
    (d) both (a) and (d)
Ans: d.

32. The conductor carries more current on the surface of in comparison to its core. This phenomenon is called
    (a) skin effect
    (b) corona
    (c) Ferranti effect
    (d) Proximity effect
Ans: a.

33. The skin effect in conductor results in
    (a) increases in its dc resistance
    (b) decrease in its ac resistance
    (c) increase in its ac resistance
    (d) None of the above
Ans: c.

34. The skin effect of a conductor reduces with the increase in
    (a) supply frequency
    (b) resistive of the conductive material
    (c) cross section of conductor
    (d) permeability of conductor material
Ans: b.

35. Skin effect in conductor is proportional to
    (a) (diameter of the conductor)$^{1/2}$
    (b) (diameter of the conductor)
    (c) (diameter of the conductor)^2
    (d) (diameter of the conductor)^4
Ans: c.

36. Increasing the frequency of transmission line will
    (a) increase shunt reactance
    (b) decrease line resistance
    (c) increase line resistance
    (d) decrease series reactance
Ans: c.

37. Skin effect exists in
    (a) cable carrying dc current
    (b) dc transmission line only
    (c) ac transmission line only
    (d) dc as well as ac transmission lines.
Ans: c.

38. Skin effect depends upon
    (a) cross section of conductor
    (b) supply frequency
    (c) permeability of conductor material
    (d) all of the above
Ans: d.

39. Which type of insulators are used on 132 kv transmission lines?
    (a) Pin type
    (b) Disc type
    (c) Shackle type
    (d) Both (a) and (c)
Ans: b.

40. Pin type insulators are generally not used for voltages exceeding
    (a) 66 kv
    (b) 33 kv
    (c) 25 kv
    (d) 11 kv
Ans: b.

41. which of the following material is not a constitutent of material used in making porcelain insulators?
    (a) Kaolin
    (b) Quartz
    (c) Silica
    (d) Felspar
Ans: c.

42. Insulators used on EHT transmission lines are made of
    (a) PVC
    (b) Porcelain
    (c) glass
    (d) Stealite
Ans: b.

43. For a 400 kv line, the number of discs in an insulator string is around
    (a) 37
    (b) 31
    (c) 25
    (d) 16
Ans: c.

44. The voltages across the various discs of a string of suspension insulators having identical discs is different due to
    (a) surface leakage currents
    (b) series capacitance
    (c) shunt capacitance to ground
    (d) series and shunt capacitances
Ans: c.

45. In a suspension type insulator the potential drop is
    (a) maximum across the lowest disc
    (b) maximum across the topmost disc
    (c) uniformly distributed over the disc
    (d) None of the above

    Ans: a.

46. The string efficiency of a high voltage line is around
    (a) 100%
    (b) 80%
    (c) 40%
    (d) 10%

    Ans: b.

47. If the frequency of a transmission system is changed from 50 Hz to 100Hz. The string efficiency will
    (a) Increase
    (b) Decrease
    (c) Remain unchanged
    (d) May increase or decrease depending on the line parameters.

    Ans: c.

48. String efficiency can be improved by
    (a) using long cross arm
    (b) using guard ring
    (c) grading the insulators
    (d) any of the above

    Ans: d.

49. The insulators may fail due to
    (a) flash-over
    (b) short-circuit
    (c) deposition of dust
    (d) any of the above

    Ans: d.

50. The ratio of puncture voltage to the flashover voltage of a line insulator is
    (a) equal to 1
    (b) lower than 1
    (c) much greater than 1
    (d) much less than 1

    Ans: c.

51. The use of guard ring
    (a) Equalizes the voltage division between insulator discs
    (b) Is unnecessary complication
    (c) Decreases string efficiency
    (d) Is necessary complication

    Ans: a.

52. Whenever the conductors are dead-ended or there is a change in the direction of transmission line, the insulators used are of the
    (a) pin type
    (b) suspension type
    (c) strain type
    (d) shackle type

    Ans: c.

53. The string efficiency of a string of suspension insulators is dependent on
    (a) size of the insulators
    (b) number of discs in the string
    (c) size of tower
    (d) none of the above

    Ans: b.

54. 100% string efficiency means
    (a) one of the insulator discs shorted
    (b) zero potential macros each disc
    (c) equal potential across each insulator disc
    (d) none of the above

    Ans: c.

55. The structure of pin insulator increases its
    (a) mechanical strength
    (b) puncture strength
    (c) flash-over voltage
    (d) thermal strength
    Ans: c.

56. A 66 kv system has string insulator having five and the earth to disc capacitance ratio of 0.10. the string efficiency will be
    (a) 89%
    (b) 75%
    (c) 67%
    (d) 55%
    Ans: c.

57. The insulators used in guy cables are
    (a) stay insulators
    (b) shackle insulators
    (c) pin type insulators
    (d) disc type insulators
    Ans: a.

58. The sag of a transmission line conductor in summer is
    (a) less than that in winter
    (b) more than that in winter
    (c) same as in winter
    (d) none of the above
    Ans: b.

59. In a transmission line, sag depends upon
    (a) span length
    (b) tension in conductor
    (c) weight of the conductor per unit length
    (d) all of the above
    Ans: d.

60. The effect of ice deposition is on conductor is to increase the
    (a) weight of the conductor
    (b) transmission losses
    (c) resistance to flow of current
    (d) skin effect
Ans: a.

61. The maximum tension in a section of overhead line conductor between two supports of unequal height occurs at
    (a) the higher support
    (b) the lower point
    (c) the mid point of the conductor
    (d) none of the above
Ans: a.

62. Stringing chart is useful
    (a) for finding the sag in the conductor
    (b) in the design of tower
    (c) in the design of insulator string
    (d) finding the distance between towers
Ans: a.

63. Hot template curves are plots of
    (a) temperature and humidity
    (b) conductor sag and span lengths
    (c) conductor weight and sag
    (d) none of the above
Ans: b.

64. The effect of wind pressure is more predominant on
    (a) insulators
    (b) transmission lines
    (c) supporting towers
    (d) none of the above
Ans: c.

65. The sag of the conductors of a transmission line is 2.5 m when the span is 250m. Now if the height of supporting tower is increased by 25%, the sag will
   (a) reduce by 25%
   (b) increase by 25%
   (c) reduce by 12.5%
   (d) remain unchanged.
Ans: d.

66. The minimum clearance of high voltage lines from ground across streets is
   (a) 3m
   (b) 5m
   (c) 6m
   (d) 8m
Ans: c.

67. Minimum horizontal clearance of a low voltage line from residential buildings must be
   (a) 0.6 m
   (b) 1.2 m
   (c) 0.9 m
   (d) 1.6 m
Ans: b.

68. If a 132 kv line passes over a residential building, the minimum vertical clearance from the roof of the building shall be
   (a) 4.57 m
   (b) 5 m
   (c) 6 m
   (d) 3 m
Ans: a.

69. The minimum clearance between 132 kv transmission line and ground is about
    (a) 6.4 m
    (b) 3.2 m
    (c) 10.5 m
    (d) 7.5 m
Ans: a.

70. For a 400 kv line, the spacing between phase conductors is around
    (a) 8 m
    (b) 11m
    (c) 14 m
    (d) 17 m
Ans: a.

71. The horizontal spacing between phase conductors of a 132 kv transmission line is about
    (a) 8 m
    (b) 6 m
    (c) 4 m
    (d) 2 m
Ans: b.

72. The difference in levels between the points of supports and the lowest point is known as
    (a) Sag
    (b) Tension
    (c) Sag template
    (d) Stringing chart
Ans: a.

73. _____ is helpful in knowing the sag and tension at any temperature
    (a) sag template
    (b) stringing chart
    (c) tension
    (d) sag
Ans: b.

74. Transmission of power by ac cables is impossible beyond
    (a) 35 – 45 km
    (b) 500 km
    (c) 300 km
    (d) 200 km
    Ans: a.

75. Sheaths are used in cables to
    (a) provide proper insulation
    (b) provide mechanical strength
    (c) provide ingress of moisture
    (d) none of the above
    Ans: c.

76. The material commonly used for sheaths of underground cable is
    (a) copper
    (b) Lead
    (c) Steel
    (d) Rubber
    Ans: b.

77. Metallic shielding is provided on underground cables to
    (a) reduce thermal resistance
    (b) reduce corona effect
    (c) control the electrostatic voltage stress
    (d) all of the above
    Ans: d.

78. Metallic shielding is provided on underground cables is usually of thickness
    (a) 0.1 – 0.8 mm
    (b) 3 – 5 mm
    (c) 10 – 15 mm
    (d) 15 – 25 mm
    Ans: b.

79. The size of the conductor used in power cables depends on
    (a) operating voltage
    (b) power factor
    (c) current to be carried
    (d) type of insulation used
Ans: c.

80. The thickness of insulation layer provided on the conductor in cables depends upon
    (a) operating voltage
    (b) Power factor
    (c) Current to be carried
    (d) Type of insulation used
Ans: a.

81. The insulating material most commonly used for power cable is
    (a) PVC
    (b) Paper
    (c) Rubber
    (d) Any of the above
Ans: b.

82. The relative permittivity of rubber is between
    (a) 0.5 and 1
    (b) 2 and 3
    (c) 5 and 8
    (d) 10 and 15
Ans: b.

83. Dielectric strength of rubber is about
    (a) 10 kv/cm
    (b) 20 kv/cm
    (c) 30 kv/cm
    (d) 100 kv/cm
Ans: c.

84. The maximum safe temperature of paper insulated cable is about
    (a) 60°C
    (b) 95°C
    (c) 135°C
    (d) 165°C
Ans: b.

85. Single core cables are usually not provided with armouring in order to
    (a) Avoid excessive loss in the armour
    (b) make the cable more flexible
    (c) make the cable non-hygroscopic
    (d) none of the above
Ans: a.

86. Multicore cables generally use
    (a) Oval shaped conductors
    (b) Sector shaped conductors
    (c) Square conductors
    (d) Either (a) or (b)
Ans: d.

87. Sulpher hexafluoride cable is insulated by
    (a) impregnated paper
    (b) polyvinyl chloride
    (c) high pressure oil
    (d) compressed gas
Ans: d.

88. The power factor of an open-ended cable can be improved by
    (a) increasing the capacitance
    (b) decreasing the capacitance
    (c) increasing the conductor resistance
    (d) increasing the insulation resistance
Ans: d.

89. To obtain the minimum value of stress in cables, the ratio (R/r) should be
    (a) 2.13
    (b) 2.718
    (c) 1.96
    (d) 1.5
Ans: b.

90. The surge impedance of a 50 miles long underground cable is 50 ohm. For a 25 miles length it will be
    (a) 25 ohm
    (b) 50 ohm
    (c) 100 ohm
    (d) None of these
Ans: b.

91. A cable carrying ac has
    (a) leakage losses only
    (b) hysteresis losses only
    (c) hysteresis and leakage losses only
    (d) hysteresis, leakage and friction losses
Ans: c.

92. The breakdown voltage of a cable depends upon
    (a) presence of moisture
    (b) operating temperature
    (c) time of application of the voltage
    (d) all of the above
Ans: d.

93. The main criterion for selection of the size of a distributor for radial distribution system is
    (a) voltage drop
    (b) corona loss
    (c) temperature rise
    (d) capital cost
Ans: a.

94. While designing the distribution to locality of one lac population with medium dense load requirement, we can employ _____
    (a) radial system
    (b) parallel system
    (c) ring main system
    (d) any of the mentioned
    Ans: a.

95. A _____ distribution system is more reliable than the _____ distribution system.
    (a) parallel, radial
    (b) parallel, ring
    (c) radial, parallel
    (d) ring, parallel
    Ans: a.

96. While designing the distribution sub stations by the designer, it is required to use the _____ for the discrete power tapping.
    (a) distributor
    (b) power transformer
    (c) distribution transformer
    (d) feeder
    Ans: a.

97. For the given distribution system the maximum voltage at the midpoint will be _____
    (a) 155.6 V
    (b) 311.12 V
    (c) 622 V
    (d) 220 V
    Ans: a.

98. Which of the following system is preferred for good efficiency and high economy in distribution system?
    (a) Single phase system
    (b) 2 phase 3 wire system

(c) 3 phase 3 wire system
(d) 3 phase 4 wire system

Ans: c.

99. The topmost conductor in hv transmission line is
    (a) B-phase conductor
    (b) Y-phase conductor
    (c) R-phase conductor
    (d) Earth conductor

Ans: d.

100. High voltage transmission lines are transposed because then
    (a) Phase voltage imbalances can be minimized
    (b) Voltage drop in the lines can be minimized
    (c) Computations of inductance becomes easier
    (d) Corona losses can be minimized

Ans: a.

Printed in the USA
CPSIA information can be obtained
at www.ICGtesting.com
CBHW020927030724
11007CB00017B/33